The Neutral Theory's Balancing Act: How Much Does Chance Matter in Speciation?

Sheena

Table of Contents

Chapter 1. Introduction

Background

Speciation is governed by both adaptive and stochastic processes. Adaptive processes have been recognized as important drivers of speciation since Darwin's seminal publication of *On the Origin of Species* (Darwin, 1859), and are regarded by many as the primary mechanism by which speciation occurs (Nosil, 2012). Biologists were more reticent to integrate stochastic processes into speciation theory, however. It took more than 70 years after *On the Origin* was published for Sewall Wright to formally introduce the concept of genetic drift, which drew heavy criticism from eminent scholars at the time (Wright, 1932). Nearly four decades later, Kimura rekindled interest in stochastic speciation processes with his contributions to the development of the Neutral Theory of Molecular Evolution, wherein he emphasized that the majority of mutations are selectively neutral and that the majority of genetic variation present in species is due to random genetic drift (Kimura, 1968). Since then, the idea that both adaptive and stochastic processes can play integral roles in the generation of biological diversity has largely been accepted, although the degree to which Neutral Theory is universally applicable has come under recent scrutiny (Kern & Hahn, 2018; but see Jensen *et al.,* 2019). Nevertheless, distinguishing the effects of adaptive vs. stochastic processes on the genome is a topic of interest for many, as doing so is essential to differentiating factors that contribute to speciation from those that do not (Sobel *et al.,* 2010). The popularity of such inferences, however, may belie the ease with which they can be made; many selective processes (*e.g.,* selective sweeps) can produce patterns of genetic diversity which are similar to expectations under genetic drift (Przeworski *et al.,* 2005).

The processes underlying speciation may also affect phylogenetic inference. A common problem encountered with phylogenetic inference is when the topologies of gene trees do not match the topology of the species tree (*i.e.,* gene tree discordance). Gene tree discordance is most often caused by two biological processes: incomplete lineage sorting and interspecific hybridization (Degnan & Rosenberg, 2009). In the case of incomplete lineage sorting (ILS), two or more lineages in a population fail to coalesce during a branching event, causing at least one of the populations to coalesce with a more distantly related population first (Degnan & Rosenberg, 2009). Under the multispecies coalescent, ILS is applied to gene trees in a species tree, and thus the failure of populations within a species to coalesce for a given gene tree would result in a topology for that gene tree which is incongruent with the species tree (Degnan & Rosenberg, 2009). ILS is an ubiquitous issue for rapidly evolving lineages (*i.e.,* rapid evolutionary radiations), because when speciation occurs rapidly, short internal branches of the tree, which shorten the window for coalescence, are more common (Whitfield & Lockhart, 2007). If the coalescent is not modeled into phylogenetic inference, or if the data violate assumptions of evolution according to a single evolutionary tree (*e.g.,* if data from incongruent gene trees are concatenated and estimated with maximum likelihood), gene tree discordance caused by ILS can mislead phylogenetic inference, potentially even leading to strong support for incorrect tree topologies (Kubatko & Degnan, 2007). Indeed, many methods for species tree inference and species delimitation are now developed with consistency under the multispecies coalescent in mind (Edwards, 2009).

In the case of interspecific hybridization, allelic variants are exchanged between different (and sometimes quite distantly related) species. An allele which is introgressed will thus more closely reflect the evolutionary history of the species under which it originally evolved. Depending on how closely related the two species in question are, the topology of the gene tree for the introgressed gene may differ greatly from the topologies of the species tree and of other gene trees present in the species (Degnan & Rosenberg, 2009). Like with ILS, many methods have now been developed which can incorporate introgression and are consistent under the coalescent model (*e.g.*, Kubatko, 2009; Folk *et al.*, 2018).

Despite these efforts, accurately estimating phylogenies for groups in which the biological processes of diversification produce extensive gene tree discordance (*e.g.*, rapid evolutionary radiations, or frequent hybridization) can be challenging. Yet, because these groups are often some of the most speciose groups across the tree of life (Schluter, 2000; Mallett *et al.*, 2016), understanding how their evolutionary histories have been shaped by such processes becomes all the more important. In my book, I focus on understanding diversification processes in the genus *Penstemon*, a large group of angiosperms which has undergone a recent and rapid evolutionary radiation. In particular, I investigate the role of both adaptive and stochastic processes in generating diversity in *Penstemon*, examining these factors across taxonomic scales: across the entire genus, within a subgenus, and within species. My analyses cover a breadth of adaptive and

stochastic mechanisms of diversification, including adaptive radiation, founder-event

speciation, and introgressive hybridization.

Study System

Penstemon Schmidel (Plantaginaceae) is the largest genus of angiosperms

endemic to North America, with nearly 300 described species (Wolfe *et al.*, 2006, 2021;

Freeman, 2019). This species-rich genus harbors an incredible amount of phenotypic and

ecological diversity, and is hypothesized to have recently undergone a rapid evolutionary

radiation (Wolfe *et al.*, 2006, 2021; Wessinger *et al.*, 2016). Species of *Penstemon* may

vary greatly from one another in their growth form, floral morphology, climatic

preferences, and soil requirements. For example, many species of *Penstemon* are found

only on particular edaphic substrates (*e.g.*, oil shale, sand dunes, and limestone outcrops),

indicating that soil may be an important factor in the diversification and distribution of

species in the genus. The phenotypic diversity, broad geographic distribution, and

evidence of repeated instances of restriction to particular ecological conditions make

Penstemon an ideal candidate in which to study both adaptive and stochastic factors

affecting biological diversification.

Molecular studies have consistently placed the subgenus *Dasanthera*, also known

as the shrubby beardtongues, as sister to the rest of *Penstemon*. Species in *Penstemon*

subgenus *Dasanthera* (hereafter, "*Dasanthera*") are found mainly in the Pacific

Northwest of North America (PNW). Five of these species mainly inhabit the Cascades and Sierra Nevada Mountains of California, Oregon, Washington, and British Columbia (*P. davidsonii, P. rupicola, P. barrettiae, P. cardwellii,* and *P. newberryi*), and three species inhabit the northern Rocky Mountains of British Columbia, Alberta, Idaho, Utah, Wyoming, and Montana (*P. ellipticus, P. lyallii,* and *P. montanus*). The final remaining species, *P. fruticosus,* is distributed in both the Cascades and northern Rocky Mountains, and in scattered mountains surrounding the Columbia Basin. *Dasanthera* species are primarily outcrossing, long-lived plants, and typically grow in a low-lying shrub or sub-shrub habit in semi-disturbed, montane environments. The apparent limitations to dispersal for *Dasanthera* species (seeds have no obvious dispersal mechanisms), is thought to contribute to their propensity to form scattered, isolated populations (Every, 1977).

Molecular analyses using ISSR data and chloroplast and nuclear loci have failed to illuminate phylogenetic relationships between *Dasanthera* species (Datwyler & Wolfe, 2004). As a result, species' relationships in *Dasanthera* remain uncertain. Complicating phylogenetic inference in *Dasanthera* is the widespread occurrence of interspecific hybridization. Hybridization is common when species are in sympatry, and there are many well-documented examples of naturally-occurring hybrid zones in *Dasanthera* (*e.g.,* Clausen *et al.,* 1940; Every, 1977; Datwyler, 2001). Hybridization is thought to be an important mechanism for diversification in *Dasanthera,* with a putative history of introgression linked to the formation of at least three taxa (*P. newberryi* var. *berryi, P. davidsonii* var. *praeteritus,* and *P. fruticosus* var. *serratus*; Every, 1977). Relatedly, introgression patterns have also been linked to the maintenance of species boundaries in

Dasanthera, whereby an elevational gradient in water availability appears responsible for high hybrid fitness (and maintenance of hybrids) at intermediate elevations but reduced fitness at more extreme (parent species) elevations in hybrid zones between *P. newberryi, P. rupicola,* and *P. davidsonii* (Datwyler, 2001; Kimball & Campbell, 2009).

Chapter 2. Phylogeographic analysis of shrubby beardtongues reveals range expansions during the Last Glacial Maximum and implicates the Klamath Mountains as a hotspot for hybridization

7

Introduction

Throughout the Quaternary Period, the Pacific Northwest of North America
(PNW) experienced dramatic shifts in climate due to the repeated expansion and retreat
of glaciers (Shafer, Cullingham, Côte, & Coltman, 2010). At the peak of the most recent
glacial expanse, known as the last glacial maximum (LGM), average temperatures in the
PNW were substantially colder than present-day (Otto-Bliesner & Brady, 2006), and
massive ice sheets rendered large regions of land inhospitable for many species (Pielou,
2008). After the LGM, conditions in the PNW became increasingly hot, reaching a
temperature maximum near the mid-Holocene period, about 6,000 years BP (Renssen *et
al.*, 2009; Wanner *et al.*, 2008).

Such dramatic fluctuations in climate undoubtedly altered species' distributions
through time. Many species may have survived in one or more pockets of suitable habitat
(*i.e.,* refugia) during these glacial cycles. Given the myriad effects that such processes
could have on the spatial distribution of contemporary genetic diversity, many studies
have focused on identifying such refugia and better understanding patterns of contraction
and expansion in response to glacial cycles (Avise, 2000; Hewitt, 2000). As a result, there

is a rich history of phylogeographic research in the PNW (Shafer *et al.*, 2010). Although the response of species to glacial cycles depends on their ecological and climatic tolerances (Hewitt, 2004; Stewart, Lister, Barnes, & Dalén, 2010), phylogeographic studies have identified several recurrent patterns of genetic differentiation across a broad range of taxa. For example, Soltis, Gitzendanner, Strenge, and Soltis (1997) described a north-south pattern of genetic differentiation in several species of plants distributed along the Cascades and Coastal Mountain ranges. In addition to this, Soltis *et al.* (1997) identified reduced genetic diversity in the northern portion of some species' ranges compared to the south, suggesting the presence of southern refugia for many species during the LGM, and the potential for multiple refugia in the coastal range and Klamath Mountains of southwestern Oregon and northwestern California. Brunsfeld, Sullivan, Soltis, and Soltis (2001) elaborated on these findings, outlining expectations for the hypotheses outlined in Soltis *et al.* (1997), and formulating new phylogeographic hypotheses for species with Cascade/Sierran distributions.

The Klamath Mountains, one of the potential refugial locations highlighted by Soltis *et al.* (1997), host a complex vegetative history owing to their old geologic age, edaphic diversity, and their ability to support mesophytic and xerophytic plant communities (Whittaker, 1961). The Klamath region has long been considered a potential refugium for plant species, as species with more northernly distributions likely invaded the Klamath Mountains during the cooler conditions of the Pleistocene, then remained there once the climate warmed again by moving to higher elevations (Smith & Sawyer, 1988; Whittaker, 1961). Indeed, the Klamath Mountains have been identified as an important geographic feature for many plant taxa, including as an area with genetic substructure or divergence (Furnier & Adams, 1986; Soltis *et al.*, 1997), as a major phylogeographic break point or area where sister species' ranges abut (Gugger, Sugita, & Cavender-Bares, 2010; Patterson & Givnish, 2003), and as a glacial refugium (Eckert, Tearse, & Hall, 2008; Kiefer, Dobeš, Sharbel, & Koch, 2009). The biogeographic importance of this region is not just limited to plants, however, as it is also a potential refugium for *Plethodon* salamanders (Pelletier, Duffield, & DeGrauw, 2011), *Anaxyrus* toads (Goebel, Ranker, Corn, & Olmstead, 2009), and *Taricha* newts (Kuchta & Tan, 2005).

Advancements in sequencing technology have given rise to an increased availability of high-throughput sequence data for use in phylogeographic studies (Garrick *et al.*, 2015), which has in turn allowed researchers to compare more complex models of demographic history (*e.g.,* Smith *et al.*, 2017). Although it has long been suggested that the expansion and contraction of species ranges in response to glacial cycles led to the formation of many hybrid zones (Hewitt, 2011; Stebbins, 1985), the use of high-throughput sequence data in phylogeographic studies has increased awareness of the prevalence and complexity of gene flow and hybridization in species' responses to climatic change (Maier, Vandergast, Ostoja, Aguilar, & Bohonak, 2019; Ruffley *et al.*, 2018; Smith & Carstens, 2020). There are numerous potential outcomes of secondary contact, but some well-known examples include lineage fusion, (Petit *et al.,* 2003), adaptive introgression (Anderson & Stebbins, 1954), and speciation via reinforcement (Butlin, 1987). In cases where lineages from separate refugia hybridize upon secondary contact, lineage fusion may increase genetic diversity (Maier *et al.*, 2019; Petit *et al.*, 2003), but could reduce genetic differentiation between lineages if contact is sustained and homogenization occurs (*e.g.,* Colella *et al.,* 2018). Hybridization after substantial climatic changes could have adaptive benefits, as novel gene combinations produced upon secondary contact could be beneficial in new, open environments (Stebbins, 1985). The onset of hybridization often does not have adaptive advantages, however. In cases where hybrids are less fit in their environment than either parent, hybridization may facilitate speciation via reinforcement (Butlin, 1987; Dufresnes *et al.,* 2020). Hybridization could also be due to the unusual circumstances surrounding the colonization of new habitats. For example, colonizers at the front-end of expansion are more likely to hybridize due to limited mate choice and a multitude of new contact points with closely related species (Currat, Ruedi, Petit, & Excoffier, 2008).

In this study, we aim to understand the response of five species of the genus *Penstemon* Schmidel (Plantaginaceae) to climatic fluctuations during the late Quaternary period, in order to better appreciate how these responses affect introgression, and in turn, phylogenetic inference. The genus *Penstemon*, commonly known as the beardtongues, is the largest genus of angiosperms endemic to North America, with nearly 300 described species (Freeman, 2019; Wolfe *et al.*, 2006). Owing its species richness to a putative adaptive radiation, *Penstemon* exhibits exceptional floral diversity, and its species occupy a wide variety of ecological niches, although in general, they prefer semi-disturbed, arid habitats (Wolfe *et al.*, 2006). Although some phylogenetic relationships between *Penstemon* species are obscured, likely due to incomplete lineage sorting (Wessinger,

Freeman, Mort, Rausher, & Hileman, 2016; Wessinger, Rausher, & Hileman, 2019), one

consistent taxonomic group has been the subgenus *Dasanthera*, which contains nine

species total, and is sister to the rest of the genus. Known colloquially as shrubby

beardtongues, members of *Penstemon* subg. *Dasanthera* are primarily outcrossing, long-

lived plants, typically persisting as low-lying subshrubs in semi-disturbed, rocky habitats.

Dispersal is apparently limited – seeds have no obvious mechanisms to facilitate wind-,

water-, or animal-mediated dispersal – and this is thought to contribute to their propensity

to form scattered, isolated populations (Every, 1977). Species in *Penstemon* subg.

Dasanthera are found mainly in mountainous areas of the PNW, but extend into

surrounding regions, including California, western Montana, northwestern Wyoming,

northern Utah, and western Nevada. Four species (*P. rupicola, P. cardwellii, P.

newberryi*, and *P. davidsonii*) have a primarily Cascades/Sierra Nevada distribution, three

species (*P. lyallii, P. ellipticus*, and *P. montanus*) have a northern Rocky Mountains

distribution, and one species (*P. fruticosus*) is distributed in both the Cascades and

northern Rocky Mountains, and in scattered mountains surrounding the Columbia Basin.

Hybridization is common in *Penstemon* subg. *Dasanthera*, and there are many well-

documented localities at which natural hybrids form, both at local scales, where persistent

backcrossing into parental species is unlikely, and at wider scales, wherever species

distributions overlap (Clausen, Keck, & Hiesey, 1940; Datwyler & Wolfe, 2004; Every,

1977). Of particular interest in this context are the Klamath Mountains. Every (1977)

identified this area as a hotspot for *Penstemon* subg. *Dasanthera* hybridization, noting the

overlap of several species' distributions, the large distance between Klamath hybrids and putative parental populations, and the decreased morphological variance in Klamath hybrids compared to known early generational hybrids elsewhere; for these reasons, Every suggested that hybridization in the Klamath Mountains was likely ancient rather than the result of recent and localized introgression. The goals of the present study are to (1) estimate relationships among species of *Penstemon* subg. *Dasanthera* (hereafter referred to as *Dasanthera*) found in the Cascades and Sierra Nevada mountains, (2) identify the location and timing of refugia for these species, and (3) identify introgressed *Dasanthera* lineages at a broad geographic scale, focusing on the Klamath Mountains.

Materials and Methods

Data generation

We collected a total of 141 samples representing *Dasanthera* species found in the Cascade and Sierra Nevada Mountains (*P. rupicola*, *P. cardwellii*, *P. newberryi*, *P. davidsonii*, and *P. fruticosus*), and 3 samples of *Penstemon montanus* var. *montanus* from Idaho for use as an outgroup. Because our goals for this study were to better understand the demographic histories of species found on the western side of the Columbia Basin, we did not include samples of *P. lyallii* and *P. ellipticus*, as these species are distributed only in the northern Rocky Mountains, east of the Columbia Basin, and are thus outside the immediate scope of this study. For this reason, we only included samples of the widespread and variable *P. fruticosus* from the Cascades Mountains, rather than including samples from the Rocky Mountains. In addition, we were unable to include samples of the rare and narrowly endemic *P. barrettiae*, which is a species of

13

conservation concern in the states of Washington and Oregon, and is found only along a roughly fifty mile stretch of the Columbia River east of Portland, OR. The majority of samples (97) were collected during the summers of 2016, 2017, and 2018. The 27 remaining samples collected by the authors were collected either in 1996 or 1999. We also included samples from 20 herbarium tissue loans from herbaria at the University of Washington and Oregon State University. These samples ranged in collection date from 1993 to 2017. In total, our sampling represents 86 unique localities, 13 of which are from the Klamath Mountains, for five of the six *Dasanthera* species and four of the five varieties present in the Cascades and Sierra Nevada Mountains, across the bulk of the range of most of these species (Table 1).

All samples collected by the authors were dried with silica gel immediately upon collection. Leaf tissues from herbarium samples were procured directly from herbarium sheets, or from additional pouches of dried material accompanying the collections when available. DNA was extracted using a modified CTAB protocol (Wolfe, 2005) and quantified using a Qubit fluorometer. We generated high-throughput sequence data using a single-enzyme digest approach to Genotyping by Sequencing (GBS), as described in Elshire *et al.* (2011). We used this approach because it is capable of generating large numbers of high-quality nuclear SNPs in the absence of a reference genome. We prepared GBS libraries using 100 nanograms of DNA from each sample and a modified version of the Elshire *et al.* (2011) protocol, which is described in Smith & Carstens (2020). We sequenced DNA libraries on an Illumina Hi-Seq 2500 using paired-end 150 bp sequencing at Novogene Corporation Inc. (Sacramento, CA). We used *ipyrad* v0.9.20

(Eaton & Overcast, 2020) for GBS data processing. We trimmed all reads to 50 bp prior to analysis and discarded reverse reads from the paired-end sequencing, because preliminary analysis indicated that doing so produced higher-quality loci with more overlap across species. We produced six different types of data sets with *ipyrad*. The first data set includes every sample in the analysis, including the outgroup. The remaining five data sets are species-specific, *i.e.*, each data set only includes samples from a single species. All data sets only include loci that are present in at least 50% of the samples in that data set, and are limited to a maximum depth of 100,000 reads. The remainder of the parameters for data processing are the default parameters for *ipyrad*.

Genetic structure and lineage tree inference

We used STRUCTURE (Pritchard, Stephens, & Donnelly, 2000) on both the inter- and intraspecific data sets to examine genetic differentiation between individuals and to assign individuals to putative genetic clusters within species. We implemented STRUCTURE with an iterative approach using the *analysis.structure* command in *ipyrad*. For each data set, we discarded the first 250,000 generations as burn-in, and ran STRUCTURE for 1,000,000 steps thereafter, performing ten replicates for each value of K tested. All other parameters were left on the default values set by the *ipyrad.analysis* toolkit. We considered values of K=1 to K=15 for the interspecific data set, and K = 1 to K = 10 for each intraspecific data set, except for *P. fruticosus*, for which we considered values of K=1 to K=5 due to its smaller sample size. We then used visualizations of log-likelihood scores and the ΔK method (Evanno, Regnaue, & Goudet, 2005) to determine the optimal value of K, and visualized results using the *ipyrad*-analysis toolkit. All downstream analyses that implement the results of STRUCTURE do so by assigning individuals to the highest percentage genetic cluster in the most likely value of K for that species.

To examine relationships among individuals, we inferred a lineage tree under the coalescent model using *SVDQuartets* (Chifman & Kubatko, 2014). For this analysis, we set lineages of *P. montanus* as an outgroup but otherwise did not partition taxa by their species identity. We used the data set which included all individuals across species boundaries and evaluated all possible quartets across 100 bootstrap replicates. After this lineage tree was constructed, we used a Quartet Sampling (QS) method (Pease, Brown, Walker, Hinchliff, & Smith, 2018) to assess phylogenetic discordance and supplement measures of branch support. We used the default parameters in the QS software, which is available on the developer's GitHub (

Species tree inference

We implemented a multispecies coalescent approach in *SVDQuartets* (Chifman & Kubatko, 2014) to infer relationships between species. For species tree inference, individuals were partitioned according to their intraspecific cluster assignment from STRUCTURE. We used the data set which included all individuals across species boundaries, set *P. montanus* as the outgroup, evaluated all possible quartets, and performed 100 bootstrap replicates. We also implemented the QS method to supplement bootstrap values on the species tree. To do so, we randomly sampled one individual from each species and performed a full quartet topology search at each node. We repeated this

approach 100 times, randomly selecting a new representative individual for each species

in each replicate. We then calculated the mean and standard deviation of concordance

measures at internal and terminal branches across replicates. For each iteration, we used

the same default settings in the QS software as used in the analysis on the lineage tree.

Species distribution models

To better understand how species' distributions have changed through time, we

built species distribution models (SDMs) for the present-day, the mid-Holocene warm

period, and the LGM. To obtain localities to build SDMs, we searched the Global

Biodiversity Information Facility (GBIF.org) for occurrence points for all of the species

in our study. We curated these results by removing duplicates, manually removing

obvious outliers, and filtering out latitude and longitude coordinates that were not precise

to at least the third decimal place. Collection localities of samples sequenced in this study

were combined with the GBIF data. This resulted in 682 occurrence points for *P.

davidsonii*, 153 occurrence points for *P. rupicola*, 189 occurrence points for *P. fruticosus*,

134 occurrence points for *P. cardwellii*, and 1419 occurrence points for *P. newberryi*. We

downloaded climate data from the WorldClim database (Hijmans, Cameron, Parra, Jones,

& Jarvis, 2005) for the present-day, the mid-Holocene, and the LGM. Data for the current

and mid-Holocene climates were at a resolution of 30 arc seconds, and LGM data were at

a resolution of 2.5 minutes. Both the LGM and mid-Holocene data sets were generated by

the CCSM4 model. We used only uncorrelated bioclimatic variables (Pearson's $r < 0.7$), and used the same variables for each species. When given a decision about which correlated variable to remove, we retained variables that we suspected would be more important for explaining species' distributions. The final variables used for SDM construction can be found in Tables 2 and 3. After locality and climate data were curated, we built SDMs with the ensemble method implemented in the R package *biomod2* (Thuiller, Georges, Engler, Breiner, & Georges, 2016), which combines the predictive outputs of different modeling approaches to better separate "true" signal from "noise" (Hao, Elith, Guillera-Arroita, & Lahoz-Monfort, 2019). We used four modeling approaches in *biomod2*: Random Forests, General Linear Models, Generalized Boosting Models, and Maximum Entropy as implemented in *Maxent* (Phillips, Anderson, & Schapire, 2006). We first cropped our raster files to -150° to -100° longitude and 35° to 65° latitude. We decided to use an extent that was larger than the range of our species of interest because it is plausible that species had distributions during the mid-Holocene and the LGM that are not confined within their current ranges. This extent also captures the distribution of the subgenus as a whole, including species that are not examined in this study. We ran five replicates per model, and implemented two different strategies to sample pseudoabsences (10,000 pseudoabsence points for each strategy): (1) the 'random' strategy, which randomly samples background pseudoabsences across the raster extent, and (2) the 'disk' strategy, which employs a spatially-dependent sampling strategy to draw pseudoabsences from a buffer zone (8km-400km) surrounding presence points. We used 80% of our occurrence points for training models, 20% for testing models, and evaluated modules using the receiver operating characteristic (ROC) score, which measures the discriminatory capability of the model. The ROC score is calculated as the area under the curve of the true positive rate vs. the false positive rate, and ranges from 0 to 1; a model with perfect predictive power has a value of 1, and a model with no predictive value has a value of 0.5. For ensemble modeling, we only included models with an ROC score > 0.85, and weighted models based on their ROC score. Ensemble models were then forecast onto current, mid-Holocene, and LGM climate conditions using *biomod2*.

Hybridization and introgression between species

We used the software *HyDe* (Blischak, Chifman, Wolfe, & Kubatko, 2018) to detect instances of hybridization and introgression between focal taxa. *HyDe* uses phylogenetic invariants to detect hybridization between two parental lineages into a hybrid lineage, can differentiate between hybrid speciation and introgression, and can distinguish whether this gene exchange has occurred in a particular individual versus the cluster as a whole. We first ran a full *HyDe* analysis, testing all possible combinations (496 total tests) of parental and hybrid lineages, and assessing statistical significance at α = 0.05 after a Bonferroni correction. To assess the degree to which clusters vs. individuals are introgressed, we subsequently implemented individual and bootstrap *HyDe* analyses on the parent-hybrid triplets that produced significant results from the full analysis.

Results

Data generation

For the combined data set, our filtering process resulted in 1,739 total retained loci, with an average of 1,523 loci per sample (±105), and 9,278 total SNPs. Average heterozygosity was estimated to be 0.0105 ± 0.0077, and our data matrix had 13.73% missing SNPs. Results for both inter- and intra-specific data sets are summarized in Table 2.

Genetic structure and lineage tree inference

STRUCTURE analysis of the interspecific data set revealed the presence of substantial admixture in most *Dasanthera* species (Figure 1). For this data set, the best K value was K = 7, with six of the clusters corresponding to each species included in the analysis, and a seventh 'ghost' cluster present in small proportions for most individuals. Most notable is the degree of admixture present in individuals morphologically identified as *P. rupicola*, *P. davidsonii*, and *P. cardwellii*. For *P. rupicola*, two individuals appear to be hybrids (admixture proportions roughly 0.5 for each contributing species) between *P. rupicola* and *P. fruticosus*. Both of these individuals were sampled north of the Columbia River, and were syntopic with *P. fruticosus*. For *P. davidsonii*, all of the individuals sampled north of the Columbia River appear to be hybrids between *P. davidsonii* and *P. fruticosus*, and two individuals south of the river exhibit substantial admixture with *P. rupicola* and *P. cardwellii*. For *P. cardwellii*, many individuals appear to exhibit introgression from *P. newberryi* and *P. rupicola*, especially those sampled from the

Klamath Mountains and the southern-most extent of the species' range in the Cascades. In general, most *P. cardwellii* individuals appear genetically heterogeneous and exhibit apparent admixture from multiple species.

Intraspecific STRUCTURE analyses revealed a geographic pattern of genetic differentiation for most species. We assigned K = 1 for *P. fruticosus*, K = 2 for *P. newberryi* and *P. davidsonii*, and K = 3 for *P. rupicola* and *P. cardwellii* (Figure 2). ΔK plots, log-likelihood estimates, and explanations for assigning the best K value for intraspecific STRUCTURE analyses can be found in Figures 3 and 4. The two genetic clusters in *P. newberryi* distinguish clusters in the Klamath Mountains from those in the Sierra Nevada Mountains. This distinction mostly corresponds to *P. newberryi* var. *berryi* (Klamath) and *P. newberryi* var. *newberryi* (Sierra). For *P. davidsonii*, the two genetic clusters correspond to the North Cascades (north of the Columbia River) and the South Cascades/Sierra Nevada Mountains. Admixture is apparent in individuals near Crater Lake, Mt. Jefferson, and in the Olympic Peninsula. For *P. rupicola*, the three genetic clusters correspond to one cluster found primarily in the South Cascades and Klamath Mountains, and two clusters found primarily in the North Cascades (hereafter differentiated as North Cascades 1 and 2; North Cascades 2 contains only two individuals). Of all the species we analyzed, a geographic pattern to genetic differentiation was the weakest in *P. rupicola*, with several thoroughly admixed individuals apparent throughout its distribution. The most heavily admixed individuals were found in the Cascades Range, south of the Columbia River, with the most genetically differentiated individuals occurring in the Klamath Mountains and North

Cascades. The three genetic clusters in *P. cardwellii* correspond to the Klamath Mountains, the Coast range and Cascades at the northern-most extent of the species' distribution, and the South Cascades. Substantial admixture is evident in three individuals from the Klamath Mountains, two individuals on the northwestern slope of Mt. Hood, and in the southern-most locality in the South Cascades.

The lineage tree inferred with *SVDQuartets* supports the monophyly of all species except *P. cardwellii* (Figure 5). *P. cardwellii*, as currently circumscribed, appears paraphyletic, although its close relationship with *P. newberryi* is well supported. Individuals of *P. cardwellii* from the Klamath Mountains (node "i", Figure 5) are supported as sister to *P. newberryi*, except for one individual (ADW 567) which is supported as sister to *P. rupicola*, albeit with poor support. Elsewhere, the relationship of *P. davidsonii* and *P. fruticosus* as sister taxa is strongly supported, and while the monophyly of *P. rupicola* is supported, the relationship of *P. rupicola* as sister to the clade containing *P. newberryi* and *P. cardwellii* is only weakly supported.

Species tree inference

The species tree (Figure 6), which inferred relationships among genetic clusters identified with the intraspecific STRUCTURE analyses, largely resembled the topology outlined in the lineage tree. As in the lineage tree, *P. cardwellii* is paraphyletic in the species tree, with individuals from the Klamath Mountains supported as sister to *P. newberryi* (bootstrap = 98/100; Quartet Concordance, $qc = 0.46$). The relationship between *P. cardwellii* from the Coast Range and Cascades is uncertain, with their placement as sister taxa supported only very weakly with bootstrap values (51/100), and

with counter-support for an alternative topology using Quartet Concordance scores ($qc =$

-0.19). The alternative topology assigns *P. cardwellii* from the South Cascades to the

clade containing *P. cardwellii* from the Klamath Mountains and *P. newberryi* (Figure 7).

In addition, both bootstrap values and *qc* scores strongly support the monophyly of *P.

newberryi* and *P. davidsonii* and the close relationship between *P. cardwellii* and *P.

newberryi*. The monophyly of *P. rupicola* is supported strongly by bootstrap values, but

only weakly with *qc* scores, and like the lineage tree, the relationship of *P. rupicola* to *P.

newberryi* and *P. cardwellii* is supported only weakly.

Species distribution models

We had generally high confidence in our SDMs, with low false discovery rates

across replicates and modeling strategies for both the 'disk' (Table 4) and 'random'

(Table 3) pseudoabsence sampling methods. Although the results for both sampling

strategies are similar, in nearly every case, average ROC scores were higher using the

'random' sampling method. To avoid issues with overfitting associated with

pseudoabsences drawn from regions much larger than a species' range (Anderson &

Raza, 2010), we discuss only the results of the 'disk' method here. The average variable

importance across replicates for each model and variable are reported in Tables 5 and 6.

SDMs for the present day identified suitable habitat that extends beyond the known

natural range of three species (*P. rupicola*, *P. fruticosus*, and *P. newberryi*) (Figure 8).

Each species had more suitable habitat during the LGM than during any other time period. Following the LGM, suitable habitat for all species shifted northward and restricted considerably, with the distribution of suitable habitat during the mid-Holocene warm period appearing similar to the distribution of suitable habitat in the present day.

Hybridization and introgression between species

Our tests of hybridization and introgression revealed substantial evidence for hybridization between species. Of the 495 total tests considered, 43 of these (8.7%) were statistically significant (Table 7). Of these significant results, 28 (65.1%) included at least one lineage from the Klamath Mountains, either as a parent or a hybrid lineage. Since only 243 of the initial 495 tests (49.1%) matched this criterion, lineages from the Klamath Mountains are overrepresented as probable players in hybridization in *Dasanthera*. A similar pattern emerges with *P. cardwellii*, as 86.0% of the significant *HyDe* tests involve this species as either a parent or hybrid, despite its inclusion in only 66.1% of the total tests considered. In total, only six of the significant *HyDe* tests do not involve *P. cardwellii*, and they identify the following clusters
as putatively introgressed: (1) *P. davidsonii* in the South Cascades and Sierra Nevada, (2) *P. rupicola* in the North Cascades (North Cascades 2), (3) *P. rupicola* in the South Cascades and Klamath Mountains, and (4) *P. newberryi* in the Sierra Nevada Mountains. Plots of the distribution of γ across bootstrap replicates for these clusters are provided in Figure 9. Using the observed site pattern frequencies, γ directly estimates the degree of admixture (Kubatko & Chifman, 2015; Blischak *et al.*, 2018) and represents the proportion of parent ancestry in an individual or cluster (*e.g.*, $\gamma = 0.5$ in a 50:50 hybrid).

24

Bootstrap resampling and hypothesis testing of particular individuals in *HyDe* allows for the detection of non-uniform introgression, which is useful in cases where introgressed genetic variation is the result of gene flow rather than hybrid speciation (Blischak *et al.,* 2018). In total, our results suggest (1) gene flow between *P. davidsonii* and *P. rupicola* in the South Cascades, (2) introgression from *P. davidsonii* into *P. newberryi* in the Sierra Nevada Mountains, (3) introgression from *P. fruticosus* into *P. rupicola* in the North Cascades (North Cascades 2), and (4) hybridization between *P. rupicola* (South Cascades and Klamath) and *P. fruticosus* to form *P. davidsonii* in the South Cascades. Hybridization tests on individuals revealed that many individuals are not hybrids, suggesting that the degree of introgression within clusters is not homogeneous (Table 8). Interestingly, *HyDe* did not identify introgression from *P. fruticosus* into *P. davidsonii* (North Cascades), despite this being apparent in STRUCTURE analyses; this might explain the identification of *P. davidsonii* in the South Cascades as a hybrid between *P. rupicola* and *P. fruticosus* (Figure 9).

Discussion

We examined genetic structure, inferred relationships between species, and reconstructed species distribution models for five PNW-distributed species in *Penstemon* subgenus *Dasanthera*. Our analyses suggest that Quaternary glacial cycles played a key role in shaping species' distributions through time, which in turn affected patterns of intraspecific genetic variation. We uncovered a prevalent north-to-south axis of genetic differentiation (Figure 2) that is consistent with patterns observed in other taxa from the region (*e.g.,* Soltis *et al.,* 1997), and our SDMs suggest that most species' ranges were larger and experienced greater connectivity during the LGM than after glacial retreat (Figure 8). The Klamath Mountains may be of particular phylogeographic importance for these species, as they likely provided suitable habitat during the LGM (Figure 8), and appear to be a hotspot for hybridization (Table 7). In turn, the large degree of hybridization in the Klamath Mountains, especially with respect to one species, *P. cardwellii,* blurs species limits and presents a challenge for inferring relationships between species (Figure 6).

The near-absence of glacial refugia

The classic paradigm regarding species' responses to glacial cycles posits that, generally, temperate species will contract their ranges during peak glacial activity, often congregating in refugia, and subsequently expand their ranges during interglacial periods (Hewitt, 2000; Hewitt, 2004). However, for the species examined in this study, we observe the inverse pattern; it appears that *Dasanthera* species had larger areas of suitable habitat during the LGM compared to the interglacial period (Figure 8). While this pattern has been observed in other species, typically it has been restricted to cold-adapted taxa (Martinet *et al.,* 2018; Stewart *et al.,* 2010) or to Neotropical systems (Leite *et al.,* 2016; Perez, Bonatelli, Moraes, & Carstens, 2016), although see Gür (2013). Our results suggest that the longstanding consensus of temperate species' responses to glacial cycles may not be as generally applicable as previously thought, especially with respect to the amount of suitable habitat during glacial expansion. We posit that for the species included in this study, large amounts of suitable habitat during the LGM resulted in greater connectivity, which facilitated both intra- and inter-specific gene flow. Subsequent range reductions during the mid-Holocene then caused population fragmentation, resulting in species' distributions that are similar to the present day. In turn, this geographic isolation may have led to genetic substructure, which was then identified with our *STRUCTURE* analyses.

An important caveat to consider is that the PNW has experienced repeated glaciation events throughout the Pleistocene, all of which undoubtedly altered species' distributions and affected the geographic context in which gene flow occurred (Hewitt, 2004; Shafer *et al.*, 2010). Given the age of the *Dasanthera* clade, estimated to have formed during the early Pleistocene roughly 1.9 MYA (Wolfe *et al.*, unpublished data), it is likely that all of the species examined in this study would have experienced several oscillations of colder periods during glaciation and warmer periods following glacial retreat. These glaciation events likely obscured more ancient and complex signals of genetic structure by facilitating pulsed hybridization events (*e.g.*, Colella *et al.*, 2018), and may help to explain the widespread occurrence of *Dasanthera* hybridization, especially with respect to *P. cardwellii*. *Penstemon cardwellii*, found in both the Oregon Coast Range as well as the inland Cascades, is the most mesic-tolerant *Dasanthera* species. Consequently, coastal *P. cardwellii* clusters may have exhibited a response to glaciation similar to other coastal species associated with PNW mesic flora, wherein species' ranges shift southward and towards the coast (*e.g.*, Smith *et al.*, 2018). Most of the suitable habitat for *P. cardwellii* during the LGM occurs within the Klamath

Mountains, and this region also hosts suitable habitat for every other *Dasanthera* species in this study (Figure 8). This makes it likely that *P. cardwellii* would have been in close contact – perhaps repeatedly – with many of its close relatives, setting the stage for a complex history of hybridization centered in the Klamath Mountains and coincident with periods of peak glacial activity.

The 'central' importance of the Klamath Mountains

Our analyses suggest the presence of suitable habitat in the Klamath Mountains during the LGM for every species included in this study (Figure 8). This, combined with our analyses implicating this region as a hotspot for interspecific hybridization (Table 7) and genetic differentiation (Figure 2), highlight the phylogeographic importance of the Klamath Mountains to *Dasanthera* species. As noted earlier, this region has been identified as an important geographic feature for many plant and animal taxa (Eckert *et al.*, 2008; Furnier & Adams, 1986; Goebel *et al.*, 2009; Gugger *et al.*, 2010; Kiefer *et al.*, 2009; Kuchta & Tan, 2005; Smith & Sawyer, 1988; Soltis *et al.*, 1997; Patterson & Givnish, 2003; Pelletier *et al.*, 2011; Whittaker, 1961). In particular, the Klamath region is thought to owe much of its diversity to its topographical complexity and its proximity to several other mountain ranges, including the Cascades and Sierra Nevada Mountains, and the coastal ranges of Oregon and California. This allows species from more northern latitudes to access the area when the climate cools, and then persist at higher elevations when conditions warm again (Smith & Sawyer, 1988; Whittaker, 1961). We hypothesize that the Klamath Mountains have served a similar role for *Dasanthera* species, and suggest that this region serves as a 'choke-point' for species' movement between the Cascades and Sierra Nevada Mountains. The following scenario could explain both the abundance of *Dasanthera* diversity in the Klamath Mountains and the relative lack of diversity in the Sierra Nevada Mountains. (1) As the climate cools, species distributed at more northerly latitudes shift their ranges southward, leading to more overlap in species' ranges, including in the Klamath Mountains. (2) Species begin to exchange genes when in sympatry, forming complex hybridization networks and blurring species limits in this region. The resulting hybrids that persist form hybrid swarms (lineage fusion), which, over time, become the predominant forms in the region. (3) As the glaciers retreat, the Klamath Mountains become less hospitable; populations that persist become geographically isolated while maintaining a genetic signature of hybridization.

Hybridization and species limits

Hybridization has undoubtedly made identifying relationships between *Dasanthera* species challenging. Previous efforts have been conducted to infer relationships between species using nuclear and chloroplast sequence data, and inter-simple sequence repeat markers (Datwyler & Wolfe, 2004). The tree produced in Datwyler and Wolfe (2004) suffers from low bootstrap support along the backbone of the tree, and its topology differs substantially from the tree presented in this study (Figure 6). This is due, at least in part, to the relative lack of parsimony-informative sites in the *ITS* and *matK* sequence data, but the prevalence of hybridization across this subgenus also almost certainly contributed to the uncertainty in species relationships observed in

29

Datwyler and Wolfe (2004). It is worth noting that Albert Every, using morphological and chemical features of *Dasanthera* taxa, suggested the same relationships between species in subgenus *Dasanthera* as identified in this study (Every, 1977). Every (1977) also noted the abundance of gene flow between species in the Klamath region, and suggested a complicated network where *P. newberryi* var. *berryi* was freely exchanging genes with *P. rupicola* and *P. cardwellii*, and that genes from *P. cardwellii* were introgressed into *P. rupicola*, which in turn introgressed into *P. davidsonii*. While the exact details of Every's hypotheses were not explicitly tested here, they do serve as a useful indicator of the complex demographic history of *Dasanthera* in the Klamath Mountains.

Our analyses were unable to determine the phylogenetic placement of *P. cardwellii* within *Dasanthera*, and suggest that *P. cardwellii*, as currently circumscribed, is paraphyletic with respect to *P. newberryi* (Figure 6). Exacerbating these challenges is the apparent proclivity of *P. cardwellii* to exchange genes with its close relatives, as evidenced by counter-support in the species tree (as opposed to low signal), and our STRUCTURE (Figure 1) and *HyDe* analyses (Table 7). For this reason, although the work presented here elucidates relationships among species distributed in the Cascades and Sierra Nevada Mountains with moderate to strong support, species limits in *Dasanthera* will need revisited, at least in the context of the clade containing *P. cardwellii* and individuals from the Klamath Mountains. Aside from *P. cardwellii*, we identified the presence of introgressed genetic material in three other species: *P. davidsonii*, *P. rupicola*, and *P. newberryi* (Figure 1; Figure 9). Generally, our results suggest that introgressed clusters identified in this study are examples of multi-generational hybridization, rather than F₁ hybrids, as not all individuals within a cluster appear introgressed, and estimates of γ tend not to be close to 0.5 (Figure 9; Table 8).

While we have explored the occurrence and geographic location of hybridization, the mechanisms underlying why it occurs – or at least, the reason why hybrids appear to persist – remains unanswered. Conversely, the question remains: what maintains species boundaries between *Dasanthera* species at all? Cross-fertilization experiments have verified that there are likely few cytogenetic barriers, if any, that prevent the formation of hybrid taxa (Every, 1977; Viehmeyer, 1958). Several studies focusing on hybrids between *P. newberryi* and *P. davidsonii* have supported this finding, and have elaborated on questions regarding hybrid fitness (Clausen *et al.*, 1940; Kimball, 2008; Kimball, Campbell, & Lessin, 2008; Kimball & Campbell, 2009). *Penstemon newberryi* and *P. davidsonii* are the only two *Dasanthera* species located in the Sierra Nevada Mountains, and they form extensive hybrid zones where their ranges overlap (Clausen *et al.*, 1940; Every, 1977). Investigations into the mechanisms controlling hybrid formation and persistence uncovered that hybrids are likely formed due to a shared pollinator community (Kimball, 2008), and that intermediate resource use and physiological tolerances likely allow hybrids to persist in intermediate environments (Kimball & Campbell, 2009). There also appear to be few cytogenetic barriers to hybrid formation between these species, although there is an apparent maternal effect to hybrid fitness with respect to elevation (Kimball *et al.*, 2008). Despite this, we have uncovered extensive hybridization elsewhere in *Dasanthera*, especially with respect to *P. cardwellii* and individuals in the Klamath Mountains, the genetic signatures of which extend well beyond this geographic region into source populations of parental species. Furthermore, we identified four additional clusters that exhibit evidence of introgression from other taxa (Figure 9; Table 8). The widespread persistence of this introgression implies that F1 hybrids are able to backcross into their parental species, making genetic differentiation via reinforcement unlikely. Therefore, while there is evidence suggesting that reduced hybrid fitness can maintain species boundaries via reinforcement in *P. davidsonii* x *P. newberryi* hybrid zones, our results indicate that reinforcement is likely not maintaining species boundaries in most other *Dasanthera* hybrid zones. In these cases, it seems more probable that species boundaries are formed and maintained when species' ranges do not overlap, limiting gene flow due to poor dispersal ability.

Tables and Figures

Table 1. Collection information for specimens included in Chapter 2.

Taxon identification, geographic coordinates, unique identifier tag and number of individuals sampled for each collection locality. Localities from the Klamath Mountains are denoted with an asterisk (*) in the 'Population ID' column.

Taxon	Latitude	Longitude	Population ID	# of Ind.
P. cardwellii	42.32805556	-123.8058333	cardwellii ADW 567 (*)	1
P. cardwellii	45.3027232	-121.8184463	cardwellii ADW 607	2
P. cardwellii	44.345946	-121.993696	cardwellii ADW 612	1
P. cardwellii	45.532477	-122.08786	cardwellii ADW 602	1
P. cardwellii	45.5383113	-122.097512	cardwellii SLD 20	1
P. cardwellii	42.0197251	-123.5241447	cardwellii SLD 11 (*)	5
P. cardwellii	44.46833333	-121.8816667	cardwellii ADW 615	1
P. cardwellii	46.314	-122.037	cardwellii SLD 111	4
P. cardwellii	44.819678	-122.088399	cardwellii_BWS_108-1-4	1
P. cardwellii	44.808767	-122.110374	cardwellii_BWS_108-5-10	1
P. cardwellii	44.401918	-121.860045	cardwellii_BWS_109	2
P. cardwellii	44.024151	-121.804755	cardwellii_BWS_110	2
P. cardwellii	43.58548	-122.663978	cardwellii_BWS_112	2
P. cardwellii	46.313721	-122.036324	cardwellii_BWS_28	3
P. cardwellii	42.626077	-123.835088	cardwellii_BWS_90 (*)	2
P. cardwellii	44.274483	-123.579297	cardwellii_BWS_91	2
P. cardwellii	45.210899	-123.758863	cardwellii_BWS_92	2
P. cardwellii	45.417722	-121.825086	cardwellii_BWS_96	2
P. cardwellii	46.29557	-122.352124	cardwellii_BWS_97	2
P. cardwellii	44.0254	-121.7957	cardwellii_OSTL_10	1
P. cardwellii	44.812	-121.7934	cardwellii_OSTL_8	1
P. dav. var. *dav.*	37.188755	-118.591939	dav. Wilson 3554	1
P. dav. var. *dav.*	44.1875	-121.88055	dav. SLD 39	1
P. dav. var. *dav.*	42.944241	-122.176547	dav._BWS_115	2
P. dav. var. *dav.*	42.895602	-122.092252	dav._BWS_116	1
P. dav. var. *dav.*	43.684535	-121.265441	dav._BWS_117	2
P. dav. var. *dav.*	44.266403	-121.787933	dav._BWS_23	3
P. dav. var. *dav.*	44.021183	-121.80011	dav._BWS_25	2
P. dav. var. *dav.*	44.023102	-121.775986	dav._BWS_27	3
P. dav. var. *dav.*	42.08102	-122.718583	dav._BWS_87 (*)	2
P. dav. var. *dav.*	45.394993	-121.658014	dav._BWS_95	2
P. dav. var. *dav.*	44.0917	-121.6467	dav._OSTL_1	1
P. dav. var. *dav.*	44.4861	-121.8386	dav._OSTL_2	1
P. dav. var. *dav.*	48.832448	-121.650946	dav._OSTL_4	1

Table 1

Table 1 continued

P. dav. var. dav.	48.734817	-120.675033	dav._UWTL_1	1
P. dav. var. dav.	42.555233	-120.791667	dav._UWTL_11	1
P. dav. var. dav.	44.0052	-121.6424	dav._UWTL_3	1
P. dav. var. dav.	43.68781	-121.25558	dav._UWTL_4	1
P. dav. var. *menz.*	48.716888	-121.078214	menz._BWS_103	2
P. dav. var. *menz.*	48.706497	-121.13966	menz._BWS_104	1
P. dav. var. *menz.*	47.50596	-123.2394	menz._UWTL_2	1
P. dav. var. *menz.*	48.52361	-120.81444	menz._UWTL_5	1
P. dav. var. *menz.*	48.9738	-120.19608	menz._UWTL_6	1
P. dav. var. *menz.*	48.9363	-120.20861	menz._UWTL_7	1
P. dav. var. menz.	48.902517	-121.466933	menz._UWTL_8	1
P. frut. var. frut.	47.411009	-121.092183	frut._BWS_102	2
P. frut. var. frut.	46.768161	-121.67451	frut._BWS_106-1-5	2
P. frut. var. frut.	46.906403	-121.578231	frut._BWS_107	2
P. frut. var. frut.	45.42266	-121.614503	frut._BWS_93	1
P. frut. var. frut.	44.509	-120.6309	frut._OSTL_15	1
P. mont. var. mont.	44.511154	-114.374046	mont._BWS_61-1-10	2
P. mont. var. idaho.	43.487	-115.4037	mont._96_0178	1
P. newb. var. berryi	40.79720833	-123.0044444	berryi ADW 1502 (*)	1
P. newb. var. berryi	40.658657	-123.219845	berryi_BWS_81 (*)	2
P. newb. var. berryi	40.585266	-123.542329	berryi_BWS_83 (*)	2
P. newb. var. berryi	41.389587	-122.993862	berryi_BWS_84-1-10 (*)	2
P. newb. var. newb.	39.282751	-120.141526	newb. SLD 77	1
P. newb. var. newb.	38.55594	-120.349923	newb. SLD 73	1
P. newb. var. newb.	38.748449	-120.051575	newb. SLD 69	1
P. newb. var. newb.	38.197559	-120.021536	newb. ADW 455	1
P. newb. var. newb.	37.181157	-118.559254	newb._BWS_70	2
P. newb. var. newb.	37.587992	-118.982346	newb._BWS_71	2
P. newb. var. newb.	38.021527	-119.26708	newb._BWS_72	2
P. newb. var. newb.	38.77672	-119.888744	newb._BWS_73	2
P. newb. var. newb.	38.615967	-119.915853	newb._BWS_74	2
P. newb. var. newb.	38.661708	-120.130671	newb._BWS_75	2
P. newb. var. newb.	38.97312	-120.094277	newb._BWS_76	2
P. newb. var. newb.	39.947764	-121.140946	newb._BWS_79	2
P. newb. var. newb.	39.880031	-121.161161	newb._BWS_80	2
P. newb. var. newb.	41.231111	-122.382305	newb._BWS_85 (*)	2
P. rupicola	42.48666667	-124.02	rupicola ADW 575 (*)	1
P. rupicola	45.5383113	-122.097512	rupicola SLD 18	1
P. rupicola	42.234281	-123.791754	rupicola SLD 14 (*)	2
P. rupicola	47.268312	-121.366167	rupicola_BWS_100	2
P. rupicola	47.411009	-121.092183	rupicola_BWS_101	3
P. rupicola	46.778033	-121.760309	rupicola_BWS_105	2
P. rupicola	46.768161	-121.67451	rupicola_BWS_106-6	1
P. rupicola	44.024151	-121.804755	rupicola_BWS_111	2

Table 1

Table 1 continued

P. rupicola	43.09002	-122.249755	rupicola_BWS_113	2
P. rupicola	41.223113	-122.37963	rupicola_BWS_86 (*)	2
P. rupicola	42.236884	-123.797584	rupicola_BWS_89 (*)	2
P. rupicola	47.436029	-121.778519	rupicola_BWS_98	2
P. rupicola	47.333277	-121.384662	rupicola_BWS_99	2
P. rupicola	44.0254	-121.7957	rupicola_OSTL_5	1
P. rupicola	44.6475	-122.0628	rupicola_OSTL_6	1
P. rupicola	42.8036	-122.2605	rupicola_OSTL_7	1
P. rupicola	47.444083	-120.94285	rupicola_UWTL_10	1
P. rupicola	47.4388	-121.7726	rupicola_UWTL_23	1

Table 2. GBS data generation statistics.

Post-processing GBS data generation statistics. All rows refer to intraspecific data sets, except for the final row, which refers to the interspecific data set. LPS = average loci per sample; SNPs = total single nucleotide polymorphisms; H_e = average heterozygosity estimates.

Species	Total loci	LPS	SNPs	% missing SNPs	H_e
P. rupicola	2403	2114 ± 213	4654	13.27%	0.0112 ± 0.0094
P. cardwellii	2434	2192 ± 112	6138	12.85%	0.0081 ± 0.0050
P. newberryi	2334	2071 ± 204	4548	12.63%	0.0090 ± 0.0059
P. davidsonii	2461	2166 ± 236	5804	12.72%	0.0150 ± 0.0086
P. fruticosus	3000	2392 ± 382	3360	19.08%	0.0080 ± 0.0087
Total (inter-species)	1739	1523 ± 105	9278	13.73%	0.0105 ± 0.0077

Table 3. Average ROC values for each model included in SDM ensemble modeling ('random' method).

Average ROC values for each model included in SDM ensemble modeling (random method). Numbers correspond to the average of ROC values across 5 replicates for each model. Models correspond to the Maximum Entropy model as implemented in *Maxent* (MAXENT.Phillips), General Linear Models (GLM), Random Forests (RF), and Generalized Boosting Models (GBM).

Species	MAXENT.Phillips	GLM	RF	GBM
P. rupicola	0.965	0.994	0.996	0.985
P. cardwellii	0.986	0.997	0.998	0.993
P. newberryi	0.976	0.998	0.998	0.997
P. davidsonii	0.905	0.994	0.993	0.993
P. fruticosus	0.958	0.955	0.959	0.960

Table 4. Average ROC values for each model included in SDM ensemble modeling (disk method).

Numbers correspond to the average of ROC values across 5 replicates for each model. Models correspond to the Maximum Entropy model as implemented in Maxent (MAXENT.Phillips), General Linear Models (GLM), Random Forests (RF), and Generalized Boosting Models (GBM).

Species	MAXENT.Phillips	GLM	RF	GBM
P. rupicola	0.841	0.957	0.962	0.960
P. cardwellii	0.937	0.970	0.968	0.967
P. newberryi	0.988	0.993	0.996	0.992
P. davidsonii	0.876	0.976	0.979	0.980
P. fruticosus	0.900	0.895	0.924	0.919

Table 5. Average variable importance for each model included in SDM ensemble modeling ('disk' method).

Models correspond to the Maximum Entropy model as implemented in Maxent (MAXENT.Phillips), General Linear Models (GLM), Random Forests (RF), and Generalized Boosting Models (GBM). Variable names correspond to bioclim variables: bio4 = Temperate Seasonality (standard deviation x 100); bio5 = Max Temperature of Warmest Month; bio8 = Mean Temperature of Wettest Quarter; bio14 = Precipitation of Driest Month; bio15 = Precipitation Seasonality (Coefficient of Variation); bio18 = Precipitation of Warmest Quarter; bio19 = Precipitation of Coldest Quarter.

P. rupicola	bio_4	bio_5	bio_8	bio_14	bio_15	bio_18	bio_19
MAXENT.Phillips	0.210	0.282	0.647	0.399	0.341	0.210	0.589
GLM	0.178	0.489	0.111	0.541	0.387	0.299	0.991
RF	0.450	0.273	0.534	0.342	0.321	0.306	0.555
GBM	0.036	0.018	0.389	0.086	0.221	0.042	0.574
P. newberryi	bio_4	bio_5	bio_8	bio_14	bio_15	bio_18	bio_19
MAXENT.Phillips	0.042	0.428	0.284	0.001	0.473	0.075	0.011
GLM	0.377	0.868	0.082	0.044	0.339	0.250	0.219
RF	0.044	0.241	0.386	0.013	0.253	0.063	0.040
GBM	0.037	0.218	0.523	0.000	0.380	0.114	0.124
P. fruticosus	bio_4	bio_5	bio_8	bio_14	bio_15	bio_18	bio_19
MAXENT.Phillips	0.114	0.042	0.773	0.548	0.400	0.135	0.220
GLM	0.307	0.165	0.152	0.440	0.367	0.305	0.056
RF	0.427	0.191	0.649	0.486	0.416	0.333	0.414
GBM	0.067	0.018	0.508	0.208	0.226	0.025	0.094
P. davidsonii	bio_4	bio_5	bio_8	bio_14	bio_15	bio_18	bio_19
MAXENT.Phillips	0.387	0.473	0.428	0.225	0.310	0.356	0.096
GLM	0.051	0.621	0.159	0.139	0.249	0.661	0.604
RF	0.114	0.158	0.315	0.071	0.270	0.177	0.049
GBM	0.105	0.159	0.475	0.004	0.419	0.065	0.003
P. cardwellii	bio_4	bio_5	bio_8	bio_14	bio_15	bio_18	bio_19
MAXENT.Phillips	0.532	0.148	0.653	0.401	0.092	0.294	0.545
GLM	0.982	0.565	0.721	0.381	0.475	0.192	0.768
RF	0.269	0.448	0.515	0.525	0.319	0.428	0.458
GBM	0.191	0.022	0.318	0.115	0.357	0.108	0.514

Table 6. Average variable importance for each model included in SDM ensemble modeling ('random' method).

Models correspond to the Maximum Entropy model as implemented in Maxent (MAXENT.Phillips), General Linear Models (GLM), Random Forests (RF), and Generalized Boosting Models (GBM). Variable names correspond to bioclim variables: bio4 = Temperate Seasonality (standard deviation x 100); bio5 = Max Temperature of Warmest Month; bio8 = Mean Temperature of Wettest Quarter; bio14 = Precipitation of Driest Month; bio15 = Precipitation Seasonality (Coefficient of Variation); bio18 = Precipitation of Warmest Quarter; bio19 = Precipitation of Coldest Quarter.

P. rupicola	bio_4	bio_5	bio_8	bio_14	bio_15	bio_18	bio_19
MAXENT.Phillips	0.183	0.066	0.750	0.270	0.517	0.190	0.735
GLM	0.595	0.007	0.307	0.473	0.125	0.199	0.860
RF	0.071	0.040	0.186	0.109	0.149	0.056	0.423
GBM	0.009	0.007	0.524	0.046	0.250	0.006	0.767
P. newberryi	bio_4	bio_5	bio_8	bio_14	bio_15	bio_18	bio_19
MAXENT.Phillips	0.177	0.250	0.265	0.096	0.372	0.541	0.038
GLM	0.381	0.291	0.184	0.074	0.136	0.640	0.156
RF	0.359	0.060	0.379	0.004	0.114	0.332	0.020
GBM	0.590	0.003	0.259	0.000	0.397	0.459	0.003
P. fruticosus	bio_4	bio_5	bio_8	bio_14	bio_15	bio_18	bio_19
MAXENT.Phillips	0.266	0.051	0.588	0.412	0.193	0.180	0.284
GLM	0.503	0.011	0.114	0.474	0.216	0.422	0.026
RF	0.203	0.080	0.515	0.199	0.172	0.241	0.191
GBM	0.332	0.007	0.579	0.112	0.060	0.055	0.138
P. davidsonii	bio_4	bio_5	bio_8	bio_14	bio_15	bio_18	bio_19
MAXENT.Phillips	0.339	0.227	0.361	0.128	0.192	0.201	0.069
GLM	0.337	0.247	0.112	0.127	0.110	0.544	0.341
RF	0.064	0.047	0.168	0.022	0.091	0.039	0.026
GBM	0.497	0.048	0.542	0.005	0.130	0.020	0.007
P. cardwellii	bio_4	bio_5	bio_8	bio_14	bio_15	bio_18	bio_19
MAXENT.Phillips	0.211	0.017	0.464	0.063	0.153	0.093	0.691
GLM	0.927	0.488	0.773	0.292	0.166	0.263	0.514
RF	0.069	0.025	0.098	0.119	0.085	0.036	0.162
GBM	0.048	0.005	0.398	0.033	0.490	0.011	0.807

Table 7. *Hyde* summary statistics.

Klamath-endemic populations include *P. newberryi* (Klamath) and *P. cardwellii* (Klamath). The chi-squared test statistic measures the difference between observed and expected proportions of significant tests including a population, with the expected number of significant tests equal to the proportion of total tests in which that population is included. Test statistics with asterisks (*) significantly deviate from expectations ($p = 0.05$, df = 1); Klamath-endemic populations and *P. cardwellii* are overrepresented, and P. rupicola is underrepresented, compared to expectations.

Population	Tests considered	Significant results	Chi-squared test statistic
Klamath-endemic populations		28 (65.1%)	4.414*
P. newberryi	243 (49.1%)	22 (51.2%)	0.073
P. davidsonii		19 (44.2%)	0.415
P. cardwellii	327 (66.1%)	37 (86.0%)	7.635*
P. rupicola		21 (48.8%)	5.719*
P. fruticosus	135 (23.7%)	8 (18.6%)	0.617
Total	495	43 (8.7%)	-

Table **8**. Summary of *HyDe* individual tests.

The column "# ind. hybrid" indicates the number of individuals from a population identified as hybrids for each test. Values for the mean and standard deviation of γ omit outlier values (individuals not identified as hybrids with a γ estimate > 2SD from the mean).

Hybrid Population	Parent 1	Parent 2	# ind. hybrid	Mean γ
P. rupicola (North Cascades 2)	*P. rupicola* (South Cascades and Klamath)	*P. fruticosus*	1/2 (50%)	0.30 ± 0.63
	P. rupicola (North Cascades 1)	*P. fruticosus*	2/2 (100%)	0.44 ± 0.35
P. davidsonii (South Cascades and Sierra)	*P. rupicola* (South Cascades and Klamath)	*P. fruticosus*	21/24 (87.5%)	0.66 ± 0.29
	P. davidsonii (North Cascades)	*P. rupicola* (South Cascades and Klamath)	21/24 (87.5%)	0.67 ± 0.17
P. rupicola (South Cascades and Klamath)	*P. rupicola* (North Cascades 1)	*P. davidsonii* (South Cascades and Sierra)	7/12 (58.3%)	0.75 ± 0.22
P. newberryi (Sierra)	*P. newberryi* (Klamath)	*P. davidsonii* (South Cascades and Sierra)	14/23 (60.9%)	0.77 ± 0.29

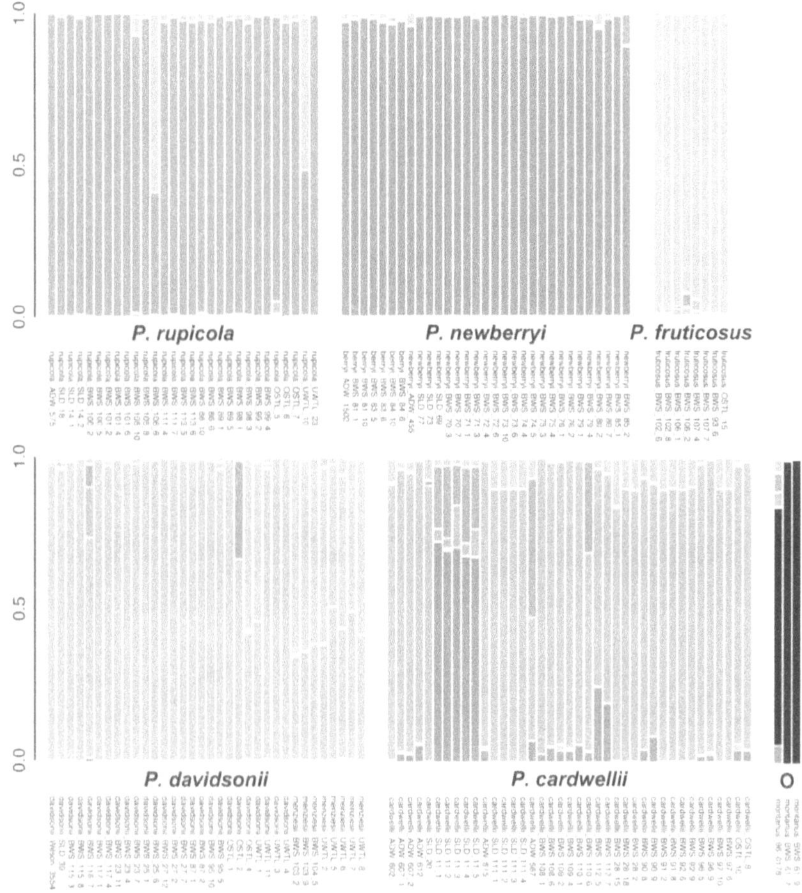

Figure 1. Plots of cluster assignment probabilities for each individual (K = 7).

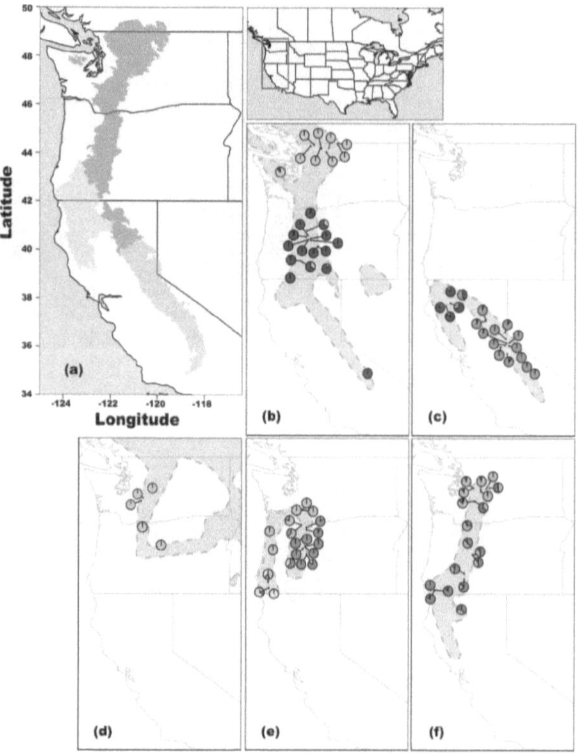

(a) Highlighted areas correspond to the Sierra Nevada (orange), Cascades and North Cascades (red), and the Klamath Mountains (blue), as defined by the United States Environmental Protection Agency. (b-f) STRUCTURE results plotted on collection localities for (b) *P. davidsonii*, (c) *P. newberryi*, (c) *P. fruticosus*, (e) *P. cardwelii*, and (f) *P. rupicola*. Colors in the pie charts correspond to the probability of membership of an individual to each of the *K* intraspecific clusters, and darkly shaded areas represent approximate species' ranges. For visualization purposes, cluster assignment probabilities of individuals from the same collection locality are averaged and represented as a single pie chart.

Figure 2. Map of the Pacific Northwest.

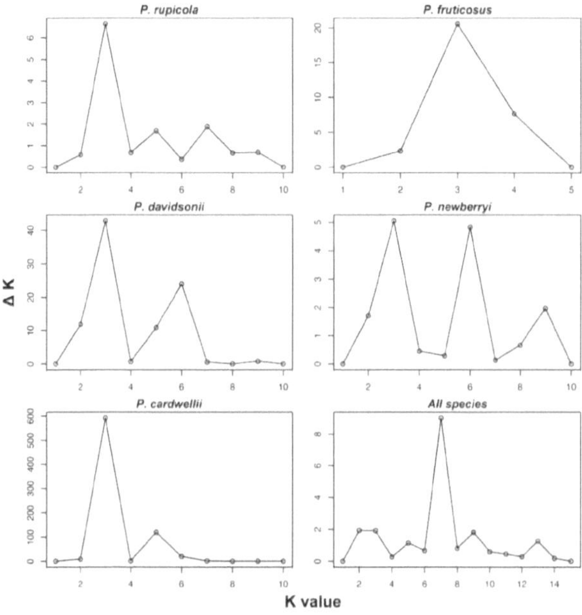

Justifications for choosing K values for each species are as follows: *Penstemon rupicola.* ΔK suggests K = 3. There are no obvious outliers in the log-likelihoods or the visualization of population assignment for this K-value, so we assigned K = 3 for *P. rupicola. Penstemon fruticosus.* ΔK suggests K = 3, but because *P. fruticosus* appears monophyletic in lineage tree analyses and is represented by only eight individuals, we assign its K-value as 1. *Penstemon davidsonii.* ΔK suggests K = 3, but K = 2 has the highest log-likelihood (Figure S2), and visualization of the population assignment distinguishes no new major lineages from K = 2 to K = 3. We therefore assigned K = 2 for *P. davidsonii. Penstemon newberryi.* ΔK suggests K values in this order: K = 6, 3, 9, and 2. However, K = 2 has the highest log-likelihood of these options (Figure S2), and visual inspection of population assignment distinguishes no new major lineages after K = 2. We therefore assigned K = 2 for *P. newberryi. Penstemon cardwellii.* ΔK suggests K = 3. There are no obvious outliers in the log-likelihoods or the visualization of population assignment for this K-value, so we assigned K = 3 for *P. cardwellii.*

Figure 3. ΔK STRUCTURE plots for each species.

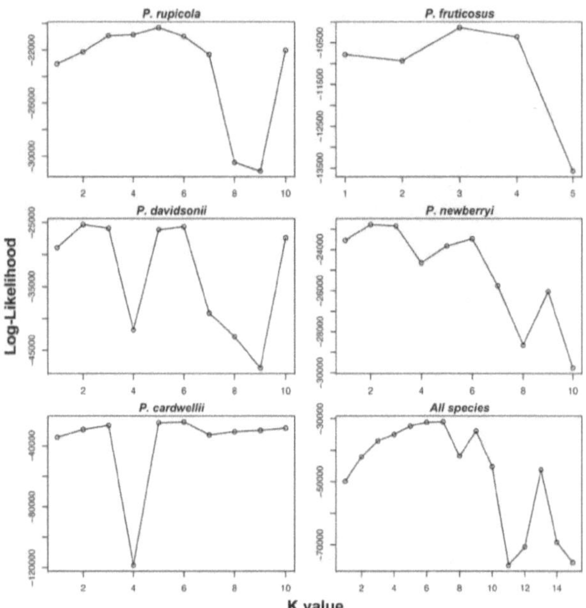

Figure 4. Log-likelihood estimates of STRUCTURE analyses for each species.

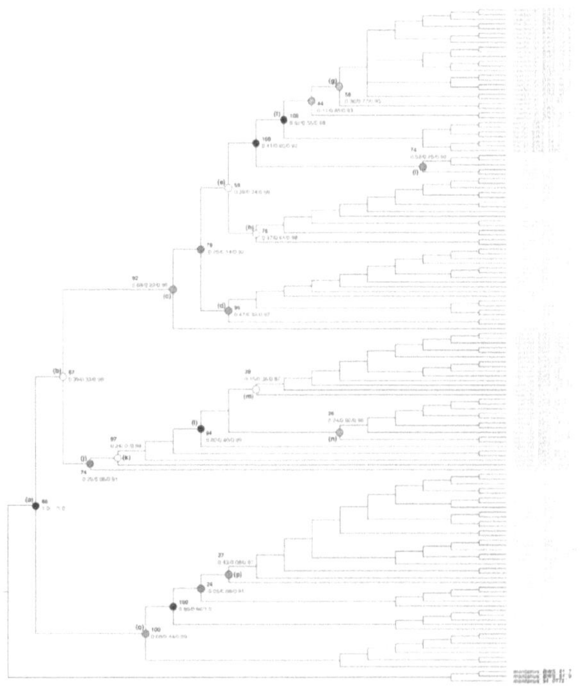

Tip labels are colored according to morphological species. At each node, bootstrap support (in bold), and quartet concordance measures (quartet concordance /differential /informativeness) are presented. Node labels correspond to the following taxonomic units of interest: (a) ingroup, (b) *P. rupicola*, *P. cardwellii*, and *P. newberryi*, (c) *P. cardwellii* and *P. newberryi*, (d) *P. cardwellii* (Coast and Cascades), (e) *P. cardwellii* (South Cascades) and *P. newberryi*, (f) *P. newberryi*, (g) *P. newberryi* (Sierra), (h) *P. cardwellii* (South Cascades, except for individual BWS 110_6), (i) *P. cardwellii* (Klamath, except for individual ADW 567), (j) *P. rupicola* + aberrant *P. cardwellii* individual (ADW 567), (k) *P. rupicola*, (l) *P. rupicola* (North Cascades 1 + South Cascades and Klamath), (m) *P. rupicola* (North Cascades 1), (n) *P. rupicola* (South Cascades and Klamath), (o) *P. fruticosus* and *P. davidsonii*, (p) *P. davidsonii* (South Cascades and Sierra).

Figure 5. Lineage tree inferred with *SVDQuartets*.

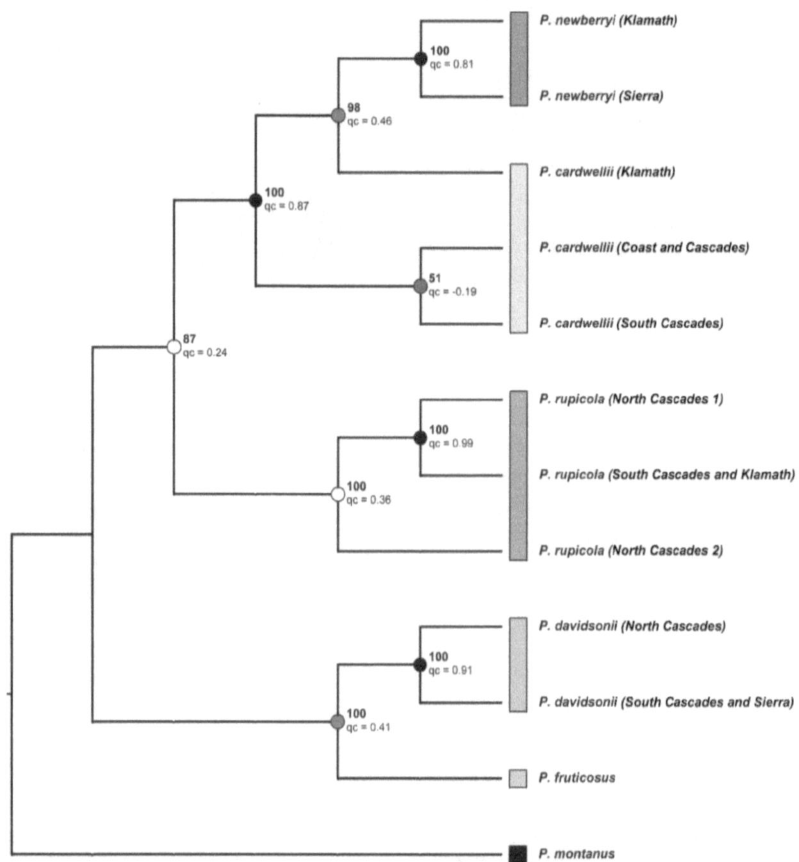

Bootstrap scores (bold) and Quartet Concordance scores (qc) are presented at the nodes. Colors at leaves indicate nominal species. Colors at nodes indicate the degree of *qc* support; black = strongly supported, grey = moderately supported, white = weakly supported, and red = counter-support.

Figure 6. Species tree constructed with *SVDQuartets*.

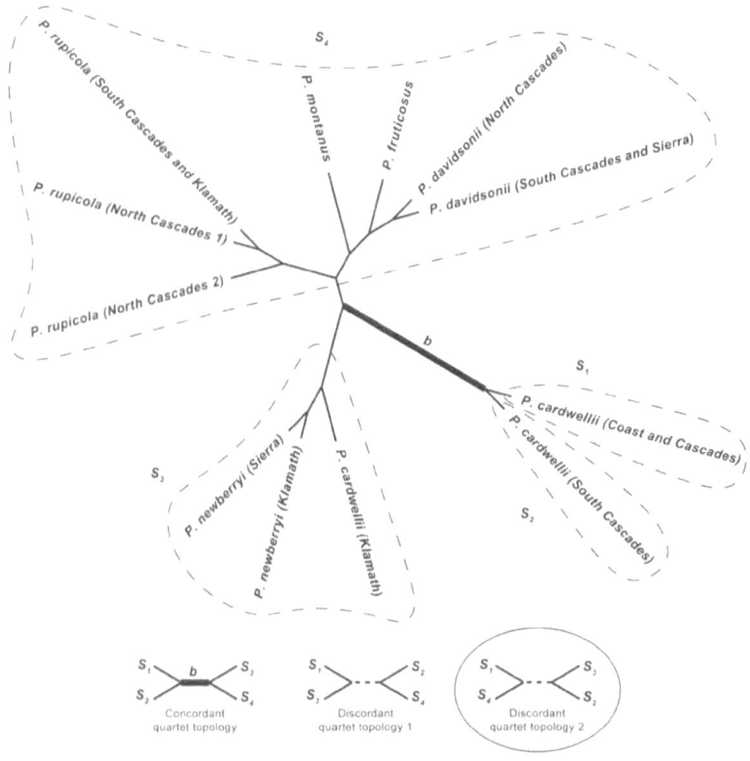

The focal branch *b* is the problematic branch in the species tree (Figure 3). Branch *b* partitions the tree into four nonoverlapping subsets of taxa, labeled S_1-S_4. These subsets can be arranged in three possible quartet topologies: the concordant topology, which is consistent with relationships in the given tree, and two discordant topologies. The QS method identified that for focal branch *b*, the second discordant quartet topology (circled below) was inferred more often than the concordant quartet topology. The layout for this figure is based on Figure 1A presented in Pease *et al.* (2018).

Figure 7. Alternative topology supported by the Quartet Sampling (QS) method.

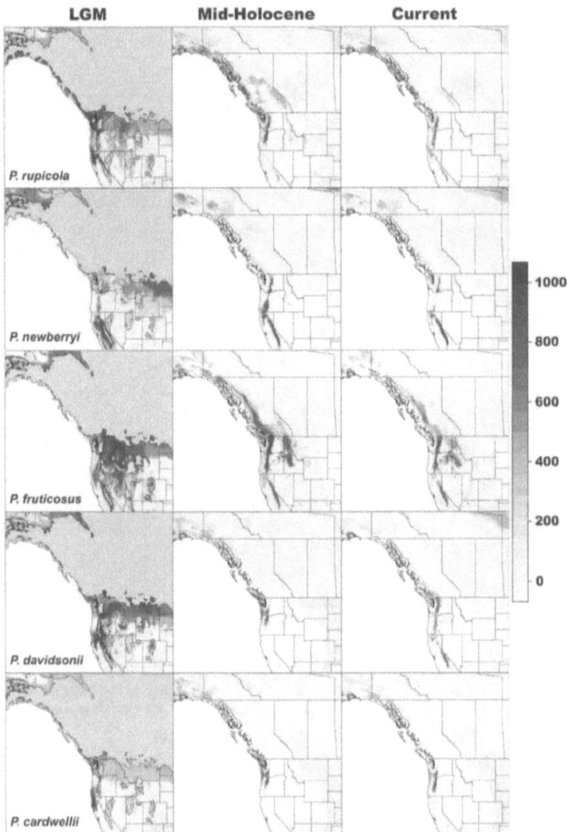

From left to right, plots indicate projections for the LGM, the mid-Holocene warm period, and the present day. Habitat suitability scores are represented by the heat map on the right; warmer colors indicate higher habitat suitability. The light blue regions in the LGM plots represent areas with glacial cover.

Figure 8. Ensemble SDMs ('disk' method) for five *Dasanthera* species.

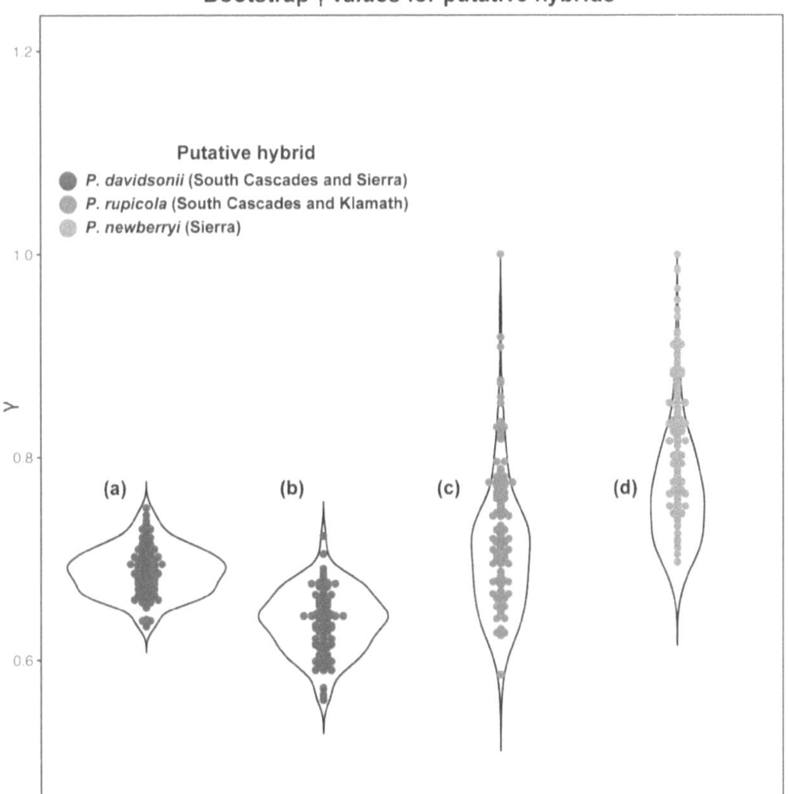

Parents for each hybrid are as follows: (a) *P. davidsonii* (North Cascades) x *P. rupicola* (South Cascades and Klamath), (b) *P. rupicola* (South Cascades and Klamath) x *P. fruticosus*, (c) *P. rupicola* (North Cascades 1) x *P. davidsonii* (South Cascades and Sierra), and (d) *P. newberryi* (Klamath) x *P. davidsonii* (South Cascades and Sierra). Note that the two tests identifying *P. rupicola* (North Cascades 2) as a hybrid are not included, as this population contains only two individuals.

Figure 9. Violin plots of *γ* across bootstrap replicates for putative hybrid populations.

51

Chapter 3. Niche and phenotype evolution in the *Penstemon* (Plantaginaceae) radiation

Introduction

Key to the study of diversification dynamics – how and why speciation occurs, and varies, across time and space – is understanding the interplay between species' phenotypes and the environment in which they live. Of particular interest to biologists in this respect are clades which exhibit great degrees of variation in the rate of net lineage diversification (speciation minus extinction) through time; identifying potential biotic and abiotic factors responsible for shifts in the rate of lineage diversification is integral to our understanding of the evolutionary processes generating biological diversity (Morlon, 2014). While there are numerous processes that can affect diversification dynamics, either ecological or otherwise, for many conceptual models, the primary sources stimulating diversification rate variation have an ecological basis (Aguilée *et al.*, 2018; Aristide & Morlon, 2019).

Perhaps the most well-known of these processes is adaptive radiation: the proliferation of ecological roles and adaptations in different species within a lineage (Givnish, 1997). Adaptive radiation does not require the rapid accumulation of species *per se*, and in fact, rapid shifts in rates of diversification ('explosive diversification') are often incorrectly synonymized with adaptive radiation (Givnish, 2015). However, it is often the case that the two processes are linked, such that the diversification into many ecologically specialized forms via adaptive radiation also confers a sharp and temporally

localized increase in the rate of lineage diversification (Gavrilets & Losos, 2009; Rundell

& Price, 2009; Moen & Morlon, 2014; Martin & Richards, 2019). Integral to adaptive

radiation theory is the concept of ecological opportunity, or the abundance of accessible

resources underused by competing taxa (Schluter, 2000). Championed as a prerequisite

for adaptive radiation by Simpson (1953), the importance of ecological opportunity stems

from the idea that an abundance of accessible resources, made available through, for

example, changes in environmental conditions, or the evolution of novel phenotypes, can

facilitate ecological specialization and ultimately the formation of new species (Simpson,

1953; Wellborn & Langerhans, 2015; Stroud & Losos, 2016). And, like adaptive

radiation, while ample ecological opportunity does not necessarily lead to an increase in

clade-wide diversification rates, bursts in lineage diversification rates attributed to

speciation driven by ecological opportunity are frequently observed in empirical studies

(*e.g.,* Burbrink & Pyron, 2010; Arakaki *et al.,* 2011; García-Navas *et al.,* 2018).

Typically, a shift in the rate of lineage diversification occurs soon after an

ecological opportunity arises, producing an 'early burst' pattern indicative of adaptation

to available niche space (Gavrilets & Losos, 2009; Yoder *et al.,* 2010; Gillespie *et al.,*

2020). This early burst pattern of lineage diversification can be accompanied by a burst in

the rate of phenotype diversification as well (Burbrink & Pyron, 2010; García-Navas *et*

al., 2018). Reduced rates of phenotype and lineage diversification are sometimes

observed subsequent to early bursts (Rundell & Price, 2009; Moen & Morlon, 2014;

Martin & Richards, 2019), and in cases where decreases in diversification rates are apparent, they tend to be attributed to density-dependent factors from increased competition between species as available niche space fills (Gavrilets & Vose, 2005; Ingram *et al.*, 2012; Aguilée *et al.*, 2018; Aristide & Morlon, 2019). However, the prevalence of diversification rate slowdowns in empirical systems, at least for phenotypic evolution, has been increasingly in question (Harmon *et al.*, 2010; Slater & Friscia, 2019; Gillespie *et al.*, 2020). Furthermore, there has been little consensus regarding the relative timing of shifts in lineage, phenotype, and niche diversification rates. While some studies have documented close associations between the timing of lineage diversification rates and rates of phenotypic (*e.g.*, Rabosky *et al.*, 2013; García-Navas *et al.*, 2018) and niche (*e.g.*, Kozak & Wiens, 2010; Title & Burns, 2015) evolution, many others report a lack of association, or a decoupling of diversification rates (*e.g.*, Adams *et al.*, 2009; Folk *et al.*, 2018; Testo & Sundue, 2018; Crouch & Ricklefs, 2019; Boucher *et al.*, 2020). This, coupled with mixed support for the ecological opportunity hypothesis at all in continental- or global-scale radiations (*e.g.*, Liedtke *et al.*, 2016; Maestri *et al.*, 2017; Folk *et al.*, 2018; García-Navas *et al.*, 2018), makes it unclear whether there is any general relationship between lineage, phenotype, and niche diversification dynamics in rapidly radiating clades, especially with respect to the relative timing of rate shifts. Such uncertainty about the relationship between rates of lineage, phenotype, and niche diversification means that expectations for a clade having potentially experienced adaptive radiation, particularly a continental radiation, are unclear. Studies of such clades, therefore, would contribute greatly to our understanding of diversification dynamics and the processes which generate biological diversity.

Here, we address this gap in our understanding of diversification dynamics through comparisons of the rate and relative timing of lineage, phenotype, and niche diversification in a continental radiation of plants, *Penstemon*. Commonly known as the beardtongues, the genus *Penstemon* has nearly 300 described species, and represents the most speciose genus of angiosperms endemic to North America (Wolfe *et al.,* 2006; Freeman, 2019). *Penstemon* exhibits an exceptional array of phenotypic diversity in floral and vegetative characters, with the degree floral diversity in particular suggesting a substantial history of selective pressure from pollinators (Straw, 1966). Species of *Penstemon* occupy a wide variety of ecological niches, with several examples of edaphic specialization (*e.g.,* sand dunes, oil shales, and calcareous soils), although most species prefer semi-disturbed, xeric habitats (Wolfe *et al.,* 2006; Wolfe *et al.,* 2021). Phylogenetic studies of *Penstemon* have revealed diversification patterns consistent with a rapid evolutionary radiation, including difficulty estimating relationships between clades early in the evolutionary history of the genus, likely due to the effects of incomplete lineage sorting (Wolfe *et al.,* 2006; Wessinger *et al.,* 2016, 2019), and an early, rapid burst in the rate of lineage diversification coincident with glacial activity during the Pleistocene (Wolfe *et al.,* 2021). Consequently, the degree of phenotypic diversity and niche specialization, and the presence of an early burst pattern of lineage diversification coincident with environmental changes at a global scale, has resulted in the hypothesis that *Penstemon* represents a continental example of a rapid adaptive radiation (Wolfe *et al.,* 2006, 2021).

We built datasets of ecological niche and phenotypic trait variables, and, using a modified version of the time-calibrated *Penstemon* phylogeny inferred in Wolfe et al. (2021), analyzed macroevolutionary patterns of lineage, phenotype, and niche diversification in *Penstemon*. Using these data, we aim to answer the following questions: (1) Do shifts in the rates of phenotype and niche diversification coincide with the shift in lineage diversification rate? (2) Do rates of lineage diversification depend on the states of phenotypic or environmental variables? (3) Does the timing of diversification coincide with potential increases in ecological opportunity due to changes in glacial activity and global temperature?

Materials and Methods

Data acquisition and processing

The phylogeny used in this study is a modified version of the time-calibrated phylogeny inferred in Wolfe et al. (2021). Briefly, this phylogeny was constructed from 43 nuclear loci generated from a targeted amplicon sequencing approach described in Blischak et al. (2014), and inferred with *RAxML* v8.2.10 (Stamatakis, 2014). Divergence times were calibrated with previously inferred divergence events in the Lamiales (Vargas *et al.,* 2014), in *BEAST* v2.6.0 (Bouckaert *et al.,* 2019) using a relaxed log-normal clock model (Drummond *et al.,* 2006); these estimates were then used as secondary calibration bounds to date the phylogeny using *treePL* (Smith & O'Meara, 2012). More details about the methodology used to produce the time-calibrated *Penstemon* phylogeny can be found in Wolfe et al. (2021).

Species occurrence data were collected for each species of *Penstemon* that was both in the constructed phylogeny and available on GBIF (gbif.org). GBIF coordinates were downloaded on June 26, 2020. For ten species (listed in Table 9), we added coordinates from the primary literature and from herbarium records accessed through the SEINet portal (swbiodiversity.org/seinet/). We retained all occurrence points accurate to at least three decimals for both latitude and longitude. We further curated our data by removing points which had identical coordinates, and occurrences which were clearly outside the species' described range, resulting in a combined total of 14,523 occurrence points. We then pruned the phylogeny to include only taxa for which there were at least three occurrence points in this curated data set.

We extracted 28 environmental layers capturing features of the climate, soil, landcover, and topology at 30s resolution. These layers included: 19 temperature and precipitation variables (bioclim variables from worldclim.org), elevation (from worldclim.org), slope, aspect, and six landcover classes from earthenv.org/landcover. Values for slope and aspect were calculated in R, using the elevation raster file. The six landcover classes correspond to the percentage landcover of evergreen/deciduous needleleaf trees, evergreen broadleaf trees, deciduous broadleaf trees, mixed/other trees, shrubs, and herbaceous vegetation. Environmental conditions for each species were extracted in R, and median values for each predictor were used in downstream analyses. Environmental data were then ordinated via phylogenetic principal components analysis (pPCA) using the 'phyl.pca' function on the correlation matrix in the R package *phytools* (Revell, 2012), forming a composite variable summarizing species' niche space.

Phenotypic traits for species were collected primarily from species descriptions in the Flora of North America (Freeman, 2019). For species not included in the Flora (mainly those endemic to Mexico), we accrued data through a combination of primary species' descriptions and measurements of specimens from herbarium collections. For these species (listed in Table 9), we downloaded images of herbarium specimens via the SEINet portal, the Steere Herbarium at the New York Botanical Gardens, and the National Autonomous University of Mexico (UNAM) Herbarium and made virtual measurements of focal phenotypic characters with *ImageJ* (Schneider *et al.,* 2012). We collected both categorical and continuous traits: the full list of traits can be found in Table 10. For continuous traits, we recorded the midpoint value, after outlier measurements were discarded. For categorical traits, we coded each descriptor sequentially, starting from zero. For species with multiple or ambiguous categorical characters, a single trait value was selected at random. We used the R package *StatMatch* (D'Orazio, 2012) to compute a phenotypic distance matrix from Gower's Distance metric (Gower, 1971), calculated with the correction proposed by Kaufman and Rousseeuw (1990). To ordinate the phenotypic data, we conducted generalized multidimensional scaling (PCOA) on the distance matrix using the 'pcoa' function in the R package *ape* (Paradis & Schliep, 2019), using the Cailliez correction (Cailliez, 1983). Because PCOA does not produce traditional loadings in the same sense that PCA does, we produced an alternative measure of variable importance for each PCOA axis. We did so by regressing the first axis of the ordinated data with each uncorrected variable and dividing the R^2 for each variable by the sum of all R^2 values, producing a relative measure of variable importance. We also constructed these values for the niche data set, to facilitate comparisons of niche and phenotype variable importance.

Ancestral state reconstruction and phylogenetic generalized least-squares

We performed ancestral state reconstruction on the ordinated phenotypic and niche data sets using the *phylopars* package (Bruggeman *et al.,* 2009) in R. We fit models of continuous trait evolution under Brownian motion (BM), Ornstein-Uhlenbeck (OU) and the 'Early Burst' model (EB), and chose the best model using the corrected Akaike information criterion (AICc; Akaike, 1973). Then, using the reconstruction from the best model, we calculated the degree of the trait shift at each node by finding the absolute difference between the reconstructed values for a focal node and its parent node. Major shifts were identified as values in the upper 95th percentile of trait shifts.

We used the *nlme* package in R to examine relationships between phenotypic and environmental variables with phylogenetic generalized least-squares (PGLS). We assessed relationships between the first axis of the ordinated phenotype data and the environmental data, including the first axis of the ordinated niche data set and each of the uncorrected environmental variables. PGLS allows estimating relationships among traits while accounting for the expected variance around the estimated slope (*i.e.,* the effect of the predictor). For each trait comparison, we tested two different models of evolution to explain correlation structures: Brownian motion (BM), and an Ornstein-Uhlenbeck model (OU). We calculated r^2 for each model using the package *rr2*, and compared model fit using AIC.

Macroevolutionary rates analyses

We used *BAMM* (Rabosky, 2014) to assess net lineage diversification through time. Prior settings were derived from the R package *BAMMtools* (Rabosky *et al.*, 2014). We ran the *BAMM* analysis for 50 million MCMC steps, in four chains, sampling every 10,000 steps. We changed the minimum clade size for shift inference equal to two (no terminal branch shifts), and set the global sampling fraction to 0.9. Otherwise, parameter values were left at their default values, or were inferred from *BAMMtools*. During post-processing, the first 10% of samples were discarded as burn-in.

We also used the *BAMM* trait model to assess the rate of both phenotype and niche evolution through time. The first axes of the ordinated phenotypic and niche data sets were used as input, and we used *BAMMtools* to generate prior settings as with the diversification analysis. For the phenotypic data, we ran the analysis for 2 billion steps, sampling every 100,000 steps, and for the niche data, we ran the analysis for 250 million steps, sampling every 10,000 steps. The first 50% and 10% of samples for the phenotypic and niche data, respectively, were discarded as burn-in during post-processing. All other settings followed those described for the *BAMM* diversification analysis. To examine variation in rates of diversification in the context of particular clades, we also generated rate-through-time plots for eleven clades, on the basis of their importance in the context of *Penstemon* taxonomy.

We used a regression approach to assess potential relationships between macroevolutionary rates and global temperature. To do so, we first obtained global temperature estimates over the past ~3.5 million years, which were derived from a temperature anomaly data set (de Boer *et al.,* 2014). We then generated a *BAMM* rates-through-time matrix for each of the three *BAMM* analyses, obtaining estimates of the average rate diversification metric at 100 discrete points in time (roughly 1000-year time slices). Values for the rate diversification metric were then paired with global temperature estimates to the nearest 1,000-year interval. We then performed regression analyses with linear, exponential, and quadratic regression, and selected the best model for each diversification metric using AIC.

We used *ES-sim* (Harvey & Rabosky, 2018) to test for relationships between trait variation and variance in lineage diversification. We performed this test for both phenotypic and niche data sets, using the first axis of each ordinated data set as input. We also ran *ES-sim* on each individual continuous phenotypic and environmental trait, to test relationships between a particular trait and rates of lineage diversification.

Results

Data acquisition and processing

Relative measures of variable importance for the ordinated phenotype and niche data sets can be found in Tables 11 and 12, respectively. The phenotypic traits with a corrected $R^2 > 0.05$ (analogous to loadings in a traditional PCA) were inflorescence type, leaf length, flower presentation, lower topography, throat expansion, plant form, and the presence/absence of basal leaves (Table 11). For the niche data set, variables meeting this criterion were mostly measures of temperature and precipitation seasonality, and the percentage of herbaceous vegetation (Table 12). Species with higher values for the niche summary statistic tend to be at higher elevations with more herbaceous landcover, more temperature seasonality (but lower minimum, maximum, and mean temperatures), and less precipitation seasonality (with less overall precipitation, but comparatively more precipitation during the dry season).

Ancestral state reconstruction and phylogenetic generalized least-squares

The best ancestral state reconstruction model for the ordinated phenotype data set was the BM model, which had an AICc score slightly more than 2 units lower than the next closest models, and comprised over half of the model weight (Table 11). Our method for identifying shifts in ancestral phenotype indicated that two of the three largest shifts in phenotype occur at the base of clades containing subgenus *Dasanthera* and subgenus *Penstemon* Section *Caespitosi* (Figure 10). These two clades also occupy a similar phenotypic space with respect to the ordinated phenotype data set. Values of the phenotype summary statistic are similar between these species, representing some of the lowest values in the genus (Figure 10). This reflects the fact that species in these clades tend to exhibit a shrub or subshrub growth form with no basal leaves, and with corolla throats expanded on the upper side. These characters were determined to be among the most important to the PCOA (Table 11), making it unsurprising that species with similarities in these traits have similar values for the phenotype summary statistic.

For the niche data set, the best model for ancestral state reconstruction was the EB model (Table 12). However, this model was nearly indistinguishable from the BM model, with AICc scores for both models very similar. Both the BM and EB models had AICc scores close to 2 units lower than the OU model, and together, they comprised > 85% of the total model weight. Some of the largest shifts in ancestral niche space were located at the base of clades containing subgenus *Penstemon* Sections *Fasciculus* and *Caespitosi* (Figure 11). This corresponds with a shift in geographic distribution for Section *Fasciculus*; the species in this clade are among the most southernly-distributed *Penstemon* species, found almost exclusively in Mexico and Guatemala. The ancestral state reconstruction reflects this shift to habitats at lower elevations with less temperature variability (but higher overall temperatures) and more precipitation variability (with more overall precipitation, except during the dry season). There are a few other clade-wide shifts (*e.g.,* at the base of Section *Caespitosi*), but these may be artifacts produced by the ancestral state reconstruction method, such that large shifts in the trait estimate at a given node does not necessarily relate to the estimate of the trait along the terminal branch. In the case of Section *Caespitosi*, a shift from a positive to a negative value occurs at the node in question, but the majority of taxa in this clade exhibit a positive value for the niche summary statistic (Figure 11).

For every trait included in the PGLS analysis, the OU model was identified as the best model, with a substantially better AIC score (< 3) in each case. This suggests that the variance seen in the phenotype summary statistic is better explained by a model incorporating both selection and drift (OU) vs. drift alone (BM). The OU model estimates the parameter a, which describes the strength by which a trait (here, the phenotype summary statistic) is pulled toward an 'optimal' value. At $a = 0$, the OU model simplifies into the BM model, and typically, estimates of a much higher than 1 indicate strong selection towards the 'optimal' trait value. Of the 29 models tested, 16 (including the niche summary statistic) were statistically significant (p-value < 0.05), suggesting significant phylogenetic signal to relationships between species' phenotype and environmental variables (Table 13). However, most of the R^2 values for these tests were low (< 0.10), indicating that, although there is significant phylogenetic signal, ultimately the environmental variables do not describe very much phenotypic variance.

Macroevolutionary rates analyses

For all three evolutionary rates analyses in *BAMM*, neither the estimates of effective sample size (ESS) or visualization of trace plots suggested failed MCMC chain convergence. For the lineage diversification analysis, the best model was one with a single rate shift (Table 14). Of the credible set of shift configurations, the single best shift configuration (proportion of posterior samples, f, = 0.88) suggests a single burst in the rate of lineage diversification about 2.5-2.0 MYA (Figure 12a). This shift includes all major clades of *Penstemon* except subgenus *Dasanthera*, *P. personatus*, *P. scapoides*, and *P. caesius*. Subsequent to this burst, the rate of diversification slowed continuously to the present day, where estimates of speciation rates are the lowest (Figure 12b). Our results very closely resemble those found in Wolfe et al. (2021) and are consistent with the hypothesis that *Penstemon* represents a recent and rapid evolutionary radiation (Wolfe *et al.*, 2006).

Analyses of rate shifts in phenotype diversification were inconclusive, with Bayes Factors continuing to increase with larger numbers of shifts, and posterior probabilities spread across shift configurations. Plots of clade-specific diversification rates show that Sections *Fasciculus* and *Spectabiles* exhibit a burst in phenotype diversification, while Section *Caespitosi* shows a slowdown of phenotype diversification (Figure 13). Analyses of niche diversification were more informative. Although Bayes Factors continued to increase as the number of shifts increased, posterior probabilities place most weight on 0-3 shifts with a posterior probability of 0.42 for zero shifts and 0.28 for a single shift affecting two species (Table 15). Plots of clade-specific diversification rates did not reveal differences between focal taxonomic groups, suggesting that our data do not point clearly to any particular shifts in niche diversification rate, although the presence of such shifts cannot be ruled out (Figure 13).

Regressions between global temperature and lineage diversification rate indicated the quadratic model as the best fit, with > 99% of the AIC model weight, and a correlation coefficient (Kendall's τ) of 0.33 (Table 16). This model suggests that lineage diversification rates were highest at intermediate global temperatures (Figure 14). The regressions between global temperature and phenotype diversification produced somewhat equivocal results; the exponential and quadratic regressions gave very similar AIC scores, with roughly equal model weights for both, and correlation coefficients (τ) were also very similar (Table 16). However, plots of both models support the same positive correlation between global temperature and phenotype diversification rates. Regressions between global temperature and niche diversification were similar to those for phenotype diversification; the exponential model is a slightly better fit than the quadratic model, and the correlation coefficients (τ) are very similar (Table 16), but plots of both models suggest a positive correlation between global temperature and niche diversification rates. We therefore present only the best models for the phenotype and niche regression analyses (Figure 14), despite only miniscule differences between the best and second-best models.

Results of all significant *ES-sim* analyses can be found in Table 17. These analyses did not reveal statistically significant relationships between variance in either the phenotype or niche summary statistic and variance in rates of lineage diversification. However, tests on unordinated phenotypic and environmental variables produced significant results for plant height, annual mean temperature (bioclim variable 1), and mean temperature of the coldest quarter (bioclim variable 11) (Table 17). These results suggest a causal relationship between mean temperature, plant height, and the rate of lineage diversification, such that taller plants, and warmer annual and winter temperatures, lead to higher rates of lineage diversification.

Discussion

We analyzed macroevolutionary patterns of lineage, phenotype, and niche diversification in *Penstemon*, a large angiosperm genus that has recently undergone a continental-scale rapid evolutionary radiation. Our results indicate that although *Penstemon* experienced an early burst of lineage diversification rates, the peak of which occurred approximately 2.5-2 MYA (Figure 12), there is no clear evidence that this was accompanied by an increase in the rate of phenotype or niche evolution. Conversely, it appears that changes in phenotype diversification rates likely operate at a smaller taxonomic scale, changing in smaller clades, rather than genus-wide, and occurring well after the initial burst in lineage diversification rates (Figure 13). We did, however, find evidence for higher rates of lineage diversification in warmer environments and in larger plants (Table 17), and it appears that, generally, rates of lineage, phenotype, and niche diversification are correlated with global temperature (Figure 14). Our results contribute to a growing body of evidence suggesting that asynchronicity in diversification rate shifts may be common, and bring into question the general applicability of expectations for diversification dynamics, which are derived mainly from studies of island systems, to studies of continental radiations.

Asynchronous shifts in rates of lineage, phenotype, and niche diversification

We observed a distinct tree-wide shift in the rate of net lineage diversification approximately 2.5-2 MYA (Figure 12) which is not evident in phenotypic or niche diversification. Rather, for niche evolution, most evidence points to no shifts in diversification rate at all (Figure 13, Table 15). For phenotype evolution, although the exact number of rate shifts is inconclusive, no single coincident shift is apparent, suggesting that changes in phenotype diversification rate occur at a smaller taxonomic scale (intra-clade vs. entire genus) and lag temporally behind the shift in lineage diversification rate (Figure 13). Similar patterns have been observed in other empirical studies. For example, lineage diversification without associated phenotypic diversification has been reported in a diverse array of taxonomic groups, including plethodontid salamanders (Adams *et al.*, 2009), birds (Crouch & Ricklefs, 2019), and ferns (Testo & Sundue, 2018). However, lags in the timing of phenotype diversification (*e.g.*, Folk *et al.*, 2018) and niche diversification (*e.g.*, McCormack *et al.*, 2010) are apparently less common. This discrepancy may be explainable if the initial burst in the rate of lineage diversification is not due to increased ecological opportunity (*i.e.*, niche-neutral), but subsequent diversification dynamics are indeed primarily density-dependent. Initial niche-neutral diversification can lead to a burst in the rate of lineage diversification without associated bursts in phenotype or niche diversification rates (Aguilée *et al.*, 2018; Folk *et al.*, 2018). If, subsequent to this burst, increases in ecological opportunity enable ecological specialization and adaptation to novel environments, ecologically-driven diversification processes (*e.g.*, niche partitioning and divergence) can begin to supersede niche-neutral processes, leading to accelerations in phenotype and niche diversification (Aguilée *et al.*, 2018). Indeed, this logic was used to explain lags in phenotype and niche diversification in the Saxifragales (Folk *et al.*, 2018). Contrary, to our findings, however, Folk et al. (2018) did not observe a slowdown in Saxifragales lineage diversification rates. This was attributed to continued formation and expansion of novel habitat, consistent with ecological opportunity caused from climate cooling and aridification after the mid-Miocene Climatic Optimum. While a slowdown is clearly evident in *Penstemon* (Figure 12), this is not in violation of theoretical expectations for density-dependent diversification dynamics; interspecific competition accelerates trait diversity, but not necessarily species richness, and in fact, competition can cause a slowdown in rates of lineage diversification even when there are no explicit ecological limits (Aristide &

Morlon, 2019). Overall, our findings are consistent with initial niche-neutral

diversification in *Penstemon*, followed by density-dependent diversification processes

that simultaneously increased phenotype diversity and led to a slowdown in lineage

diversification rates.

Geography, ecological opportunity, and the primary mode of speciation

Although the initial burst in lineage diversification rates in *Penstemon* is

consistent with niche-neutral diversification, we did also find evidence of a link between

lineage diversification rate and ecological opportunity in the form of global temperature

(Figure 14; Table 16). Relatedly, we found evidence for higher rates of lineage

diversification in warmer environments and in larger plants (Table 17). If the initial burst

in lineage diversification is primarily niche-neutral, why do there still appear to be

ecological components to lineage diversification rates in *Penstemon*?

The timing of the burst in *Penstemon* lineage diversification rate corresponds to

the onset of the Quaternary Period approximately 2.58 MYA (Head & Gibbard, 2015).

With the Quaternary Period came severe environmental changes, including the

development of ice sheets and the intensification of glaciation in the Northern

Hemisphere (Head & Gibbard, 2015). A natural conclusion, then, would be to associate

the shift in the rate of lineage diversification in *Penstemon* with the increased ecological

opportunity brought about by the cooling of the climate and the appearance of novel

niches due to Quaternary glaciation. However, it is perhaps more likely that these relationships are a byproduct of the timing of the *Penstemon* radiation, and not a cause,

per se. While ecological opportunity does increase during phases of rapid environmental

change (Wellborn & Langerhans, 2015), and undoubtedly, Quaternary glaciation had

profound effects on the geographic distributions and genomes of species (Hewitt, 2000),

the lack of corresponding phenotype and niche diversification in *Penstemon* suggests that

although initial lineage diversification coincided with large-scale environmental change,

speciation was still mostly occurring in environmental conditions similar to ancestral

habitats. In this scenario, initial lineage diversification in *Penstemon* may have been a

primarily geographic phenomenon: spurred by environmental changes, but only because

those changes resulted in the formation of abundant habitat similar to habitats in which

Penstemon species were already found.

Evidence from species distribution models in *Penstemon* subgenus *Dasanthera*

lends some support for this scenario (Stone & Wolfe, 2020). For subgenus *Dasanthera*,

which is sister to the rest of *Penstemon* and is not included in the lineage diversification

rate shift, periods of peak glacial expanse likely brought about large increases in suitable

habitat, particularly in the Columbia Basin and the Great Basin (Stone & Wolfe, 2020).

These geographic regions, while ecologically similar to ancestral habitats, are now

disjunct from current species' distributions, although a few narrowly endemic taxa (*i.e.,*

P. davidsonii var. *praeteritus* and *P. fruticosus* var. *serratus*) still persist in isolated

mountain ranges. A similar scenario for other *Penstemon* species, then, seems plausible,

especially given the abundance of narrowly endemic species (most of which,

consequently, are found in the Great Basin) and the lack of signal for shifts in niche diversification in our analyses (Table 15; Figure 13). Indeed, a strong geographic component to speciation has been confirmed to be an important driver of diversification in *Penstemon*, with associations between founder-event speciation events and elevated rates of net lineage diversification pointing to a likely mechanism for the rapid radiation of the genus (Wolfe *et al.*, 2021). The importance of these founder-event speciation events suggests that the dispersal of *Penstemon* species to previously unoccupied areas was key to the diversification of the genus (Wolfe *et al.*, 2021). Likewise, the timing of these diversification bursts, coincident with pulses of glaciation during the Pleistocene, likely reflect repeated migration events into the Intermountain Region (Cronquist, 1978), and are consistent with expectations for species' responses to ecological opportunity in an adaptive radiation (Simpson, 1953; Wolfe *et al.*, 2021). With respect to the mechanism of speciation, a primarily niche-neutral process with a strong role of geography could be responsible for such patterns of lineage diversification. One possible example is the process of 'budding' speciation (*sensu* Anacker & Strauss, 2014), whereby larger-ranged progenitor species give rise to narrow-ranged species, often causing closely related species to be sympatric. Density-dependent processes could then follow this initial niche-neutral diversification, resulting in a slowdown of the overall lineage diversification process, increasing clade-specific rates of phenotype diversification, and generating associations between species' phenotypes and the environment in which they live (Table 13). Future studies on the geography of speciation in *Penstemon*, especially regarding asymmetry in the size of species' distributions and the degree of ecological and phenotypic overlap between close relatives, would be able to assess the support for this hypothesis.

Despite the apparent importance of niche-neutral processes in generating diversity in *Penstemon*, the influence of adaptive processes, particularly those related to floral evolution and the selective pressures induced by pollinators, should not be ignored. While the formation of polyploid taxa is certainly an important diversification mechanism (*e.g.,* Keck, 1932, 1945), with virtually every *Penstemon* researcher since Pennell acknowledging the incredible diversity of floral shapes, sizes, and colors, it is clear that pollinator pressure has played a substantial role in the evolutionary history of this genus (Pennell, 1935; Straw, 1966; Wilson *et al.,* 2004; Wolfe *et al.,* 2006; Wessinger *et al.,* 2019). Perhaps unsurprisingly, then, many studies have provided evidence of adaptation to different pollinators in *Penstemon* (*e.g.,* Straw, 1956, 1963; Crosswhite & Crosswhite, 1966), with transitions from insect- to hummingbird pollination especially common (Straw, 1963; Wilson *et al.,* 2007; Wolfe *et al.,* 2006, 2021). It is for this reason that the most likely density-dependent processes driving post-dispersal speciation to unoccupied niches involve adaptations to different pollinator types. This speciation process – relatively sluggish, compared to the multitude of founder-event speciation events associated with dispersal to unoccupied niches – would account for both the slowdown in lineage diversification rates (Figure 12) and clade-specific bursts in phenotype diversification rates (Figure 13) we have presented here. Supporting this claim, transitions from insect- to hummingbird pollination, while relatively common in

Penstemon, have been associated with slowdowns in rates of diversification (Wessinger *et al.*, 2019).

In light of this, it is important to consider the limitations of the analyses we have conducted in this study. By virtue of reducing the dimensionality of our data set into a single summary statistic, we capture less of the variance between species than is explained by the data set as a whole. While reducing the complexity of the data in this way is necessary to perform many of the analyses we have presented here, it also creates a dilemma we must reconcile; it is possible – perhaps even likely – that the summary statistics employed in this study, despite explaining a large proportion of variance, may not be biologically meaningful summarizations of phenotypic and niche diversity in *Penstemon*. Future comparative studies interested in explaining diversification patterns in *Penstemon* may therefore benefit from generating a phenotypic data set more acutely focused on explaining the diversity of floral types evident in the genus.

Tables and Figures

Table 9. Species with data from additional sources.

Species	Occurrence points	Herbarium measurements
P. amphorellae	n	y
P. campanulatus	n	y
P. coriaceus	y	n
P. fasciculatus	n	y
P. gentianoides	n	y
P. goodrichii	y	n
P. hartwegii	n	y
P. hesperius	y	n
P. hidalgensis	n	y
P. imberbis	n	y
P. isophyllus	n	y
P. leonensis	n	y
P. miniatus	n	y
P. occiduus	n	y
P. pahutensis	y	n
P. parvus	y	n
P. penlandii	y	n
P. plagapineus	n	y
P. reidmoranii	y	n
P. roseus	n	y
P. saltarius	n	y
P. smallii	y	n
P. versicolor	y	n
P. vizcainensis	n	y
P. wislizeni	y	y

Table 10. Phenotypic traits collected.

Trait
Height (cm)
Leaf length (mm)
Leaf width (mm)
Corolla length (mm)
Corolla tube length (mm)
Corolla Throat diameter (mm)
Staminode length (mm)
Style length (mm)
Form
Basal leaves
Stamen location
Staminode location
Lifespan
Inflorescence
Flower presentation
Flower shape
Corolla throat expansion
Corolla lower topography
Anther dehiscence pattern

Table 11. Model weights for Ancestral State Reconstruction and variable importance (phenotype).

Model Testing			
	BM	OU	EB
AICc	-233.499	-231.495	-231.498
AICc weights	0.576	0.212	0.212
Variable Importance			
Variable	Intercept	R^2	R^2 corrected
Staminode length (mm)	0.029	0.002	0.001
Tube length (mm)	0.055	0.004	0.001
Throat diameter (mm)	0.050	0.006	0.002
Flower shape	-0.049	0.015	0.005
Stamen location	-0.033	0.016	0.005
Height (cm)	-0.075	0.037	0.011
Corolla length (mm)	0.157	0.041	0.012
Staminode location	-0.044	0.050	0.015
Style length (mm)	0.207	0.094	0.028
Leaf width (mm)	-0.122	0.120	0.036
Lifespan	0.259	0.130	0.039
Anther dehiscence	-0.240	0.158	0.047
Inflorescence	-0.035	0.213	0.064
Leaf length (mm)	-0.276	0.225	0.067
Flower presentation	0.229	0.242	0.072
Lower topography	-0.046	0.336	0.100
Throat expansion	0.094	0.435	0.130
Form	0.110	0.594	0.177
Basal leaves	0.141	0.633	0.189

Table 12. Model weights for Ancestral State Reconstruction and variable importance (niche).

Model Testing			
	BM	OU	EB
AICc	1402.215	1404.219	1401.994
AICc weights	0.403	0.148	0.450

Variable Importance			
Variable	Intercept	R^2	R^2 corrected
Slope	1.016	0.001	0.000
Aspect	1.236	0.001	0.000
% shrubs	0.895	0.002	0.000
Precip. of driest quarter	0.299	0.006	0.001
% deciduous broadleaf trees	0.844	0.012	0.002
% evergreen broadleaf trees	0.783	0.014	0.002
Precip. of driest month	-0.057	0.025	0.003
Mean temp. of wettest quarter	1.777	0.038	0.005
Mean diurnal temp. range	-7.644	0.049	0.006
% mixed/other trees	1.757	0.050	0.006
% needleleaf trees	1.564	0.051	0.007
Max temp. of warmest month	8.354	0.062	0.008
Precip. of warmest quarter	2.654	0.108	0.014
Mean temp. of warmest quarter	8.376	0.168	0.022
elevation	-3.411	0.171	0.022
Precip. of coldest quarter	2.562	0.204	0.026
Annual precip.	4.372	0.297	0.038
Mean temp. of driest quarter	3.815	0.433	0.056
Precip. of wettest quarter	4.465	0.452	0.058
% herbaceous vegetation	-2.176	0.458	0.059
Precip. of wettest month	4.683	0.472	0.061
Isothermality	18.126	0.506	0.065
Annual mean temp.	6.886	0.564	0.073
Precip. seasonality	7.004	0.583	0.075

Table 12

Table 12 continued

Temp. seasonality	-13.212	0.652	0.084
Annual temp. range	-22.126	0.696	0.090
Mean temp. of coldest quarter	0.926	0.809	0.104
Min temp. of coldest month	-3.869	0.865	0.112

Table 13. Significant PGLS results.

The parameter a corresponds to the strength of return to a theoretical optimum trait value.

Variable	p-value	R^2 Predictors	Slope	a
Niche summary statistic	0.0289	0.0354	0.0083	1.6629
Slope	0.0000	0.0857	-0.0363	2.2172
% needleleaf trees	0.0000	0.0961	-0.0030	1.5378
% shrubs	0.0189	0.0309	0.0014	1.8572
% herbaceous vegetation	0.0051	0.0435	0.0027	1.7110
Mean diurnal temp. range	0.0082	0.0480	0.0247	1.5561
Temp. seasonality	0.0333	0.0287	0.0002	1.8189
Max temp. of warmest month	0.0119	0.0346	0.0101	1.8306
Temp. annual range	0.0053	0.0440	0.0080	1.7038
Mean temp of wettest quarter	0.0009	0.0639	0.0062	1.5232
Mean temp. of driest quarter	0.0341	0.0324	-0.0039	1.6506
Mean temp. of warmest quarter	0.0275	0.0291	0.0083	1.8493
Annual precip.	0.0000	0.0882	-0.0002	1.5055
Precip. of wettest month	0.0001	0.0775	-0.0010	1.5333
Precip. of wettest quarter	0.0001	0.0790	-0.0004	1.5357
Precip. of coldest quarter	0.0000	0.1162	-0.0005	1.4411

Table 14. Bayes Factors and posterior probabilities of *BAMM* lineage diversification.

Bayes Factors			
Number of shifts	1	2	3
1	-	6.01	34.41
2	0.17	-	5.72
3	0.03	0.17	-
Posterior Probabilities	0.92	0.08	0.01

Table 15. Bayes Factors and posterior probabilities of *BAMM* niche diversification.

Bayes Factors increased for up to 19 shifts in niche diversification rate. Presented here are Bayes Factors for 0-5 shifts.

Bayes Factors						
Number of shifts	0	1	2	3	4	5
0	-	0.49	0.33	0.24	0.19	0.16
1	2.05	-	0.67	0.49	0.39	0.32
2	3.06	1.50	-	0.73	0.58	0.48
3	4.22	2.06	1.38	-	0.80	0.66
4	5.28	2.58	1.72	1.25	-	0.83
5	6.35	3.10	2.07	1.51	1.20	-
6	7.99	3.90	2.61	1.90	1.51	1.26
7	8.33	4.07	2.72	1.98	1.58	1.31
8	9.53	4.65	3.11	2.26	1.80	1.50
9	10.83	5.29	3.53	2.57	2.05	1.70
10	8.95	4.37	2.92	2.12	1.69	1.41
11	13.33	6.51	4.35	3.16	2.52	2.10
12	14.99	7.32	4.89	3.56	2.84	2.36
13	18.32	8.95	5.98	4.35	3.47	2.89
14	16.66	8.13	5.44	3.95	3.15	2.62
15	19.99	9.76	6.52	4.74	3.78	3.15
16	26.65	13.01	8.70	6.32	5.04	4.20
17	26.65	13.01	8.70	6.32	5.04	4.20
18	53.30	26.02	17.40	12.64	10.09	8.39
19	426.42	208.17	139.18	101.14	80.71	67.15
Posterior Probabilities	0.24	0.25	0.19	0.13	0.08	0.05

Table 16. Summary of regressions between global temperature and diversification rates.

Model	p-value	Correlation coefficient	AIC	AIC weight
Lineage				
linear	0.768	0.030	362.000	0.000
quadratic	0.000	0.333	340.921	1.000
Phenotype				
linear	0.000	0.709	-754.622	0.000
exponential	0.000	0.532	-779.731	0.540
quadratic	0.000	0.540	-779.411	0.460
Niche				
linear	0.000	0.751	501.883	0.001
exponential	0.000	0.592	489.209	0.553
quadratic	0.000	0.594	489.637	0.446

Table 17. Significant *ES-sim* results.

Table 4. Significant *ES-sim* results

Variable	rho	R^2	p-value
Annual mean temp.	0.307	0.094	0.039
Mean temp. of coldest quarter	0.296	0.088	0.047
Height (cm)	0.351	0.123	0.015

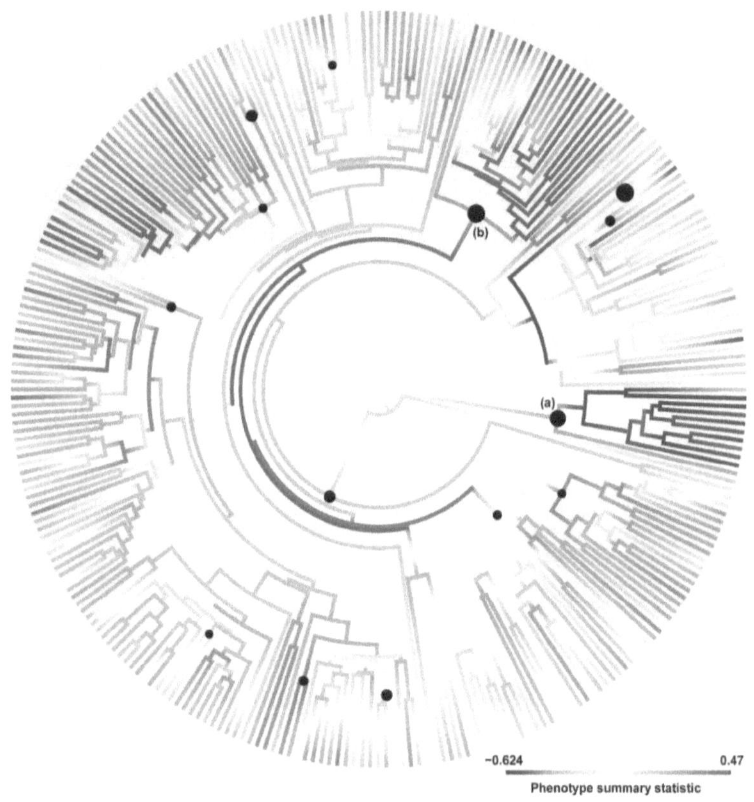

Labeled clades contain (a) subgenus *Dasanthera* and (b) Section *Caespitosi*. Trait values represent scores of the first PC axis of ordinated phenotype data; cooler values indicate (generally) short plants with a shrub or subshrub growth form, no basal leaves, and with corolla throats expanded on the upper side. Black circles at nodes indicate the largest shifts (top 5%) in ancestral state estimates from the preceding node. Circles are scaled to the size of the shift in trait value.

Figure 10. Ancestral state reconstruction of the phenotype summary statistic.

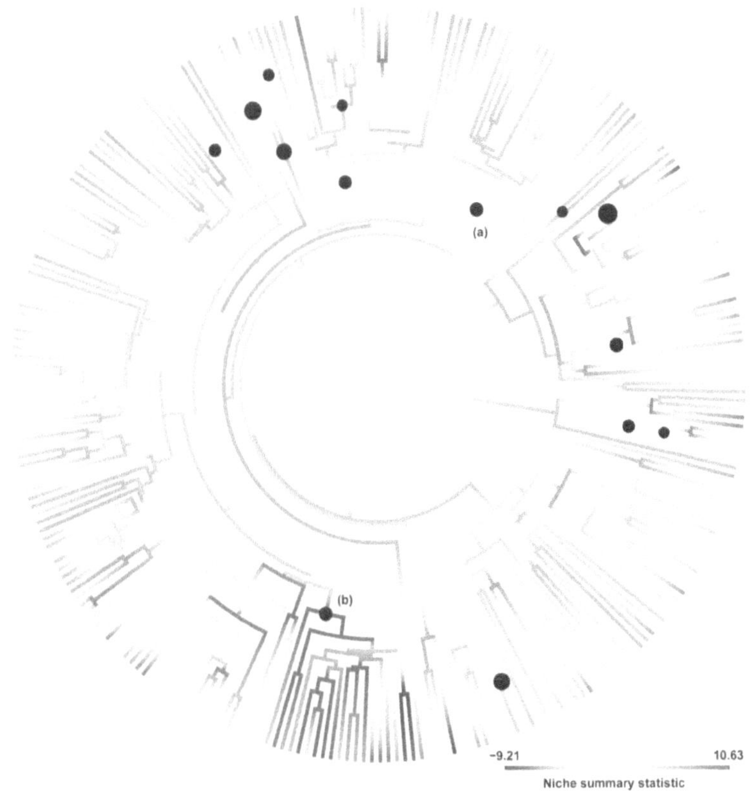

Labeled clades contain Sections (a) *Caespitosi* and (b) *Fasciculus*. Trait values represent scores of the first PC axis of ordinated niche data; warmer values indicate (generally) higher elevations with more herbaceous landcover, more temperature seasonality (but lower minimum, maximum, and mean temperatures), and less precipitation seasonality (with less overall precipitation, but comparatively more precipitation during the dry season). Black circles at nodes indicate the largest shifts (top 5%) in ancestral state estimates from the preceding node. Circles are scaled to the size of the shift in trait value.

Figure 11. Ancestral state reconstruction of the niche summary statistic.

85

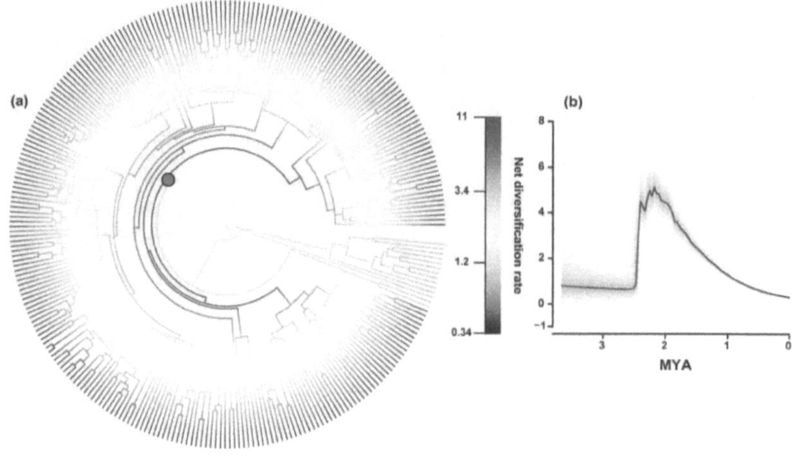

(a) *BAMM* suggested a single shift in the rate of lineage diversification early in the evolutionary history of genus (red dot), followed by a subsequent slowdown of rates of lineage diversification. (b) Lineage diversification rates increased sharply between 2.5-2 MYA, declining steadily thereafter. Both findings are consistent with results from Wolfe *et al.*, (2021).

Figure 12. Best shift configuration and RTT plot of lineage diversification rates.

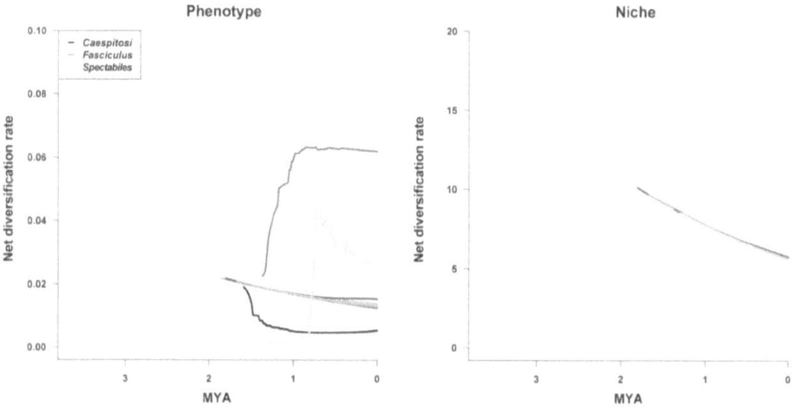

Clades differing from the general pattern of decreasing diversification rates are indicated in the top-left corner.

Figure 13. Clade-specific rates of phenotype and niche DTT.

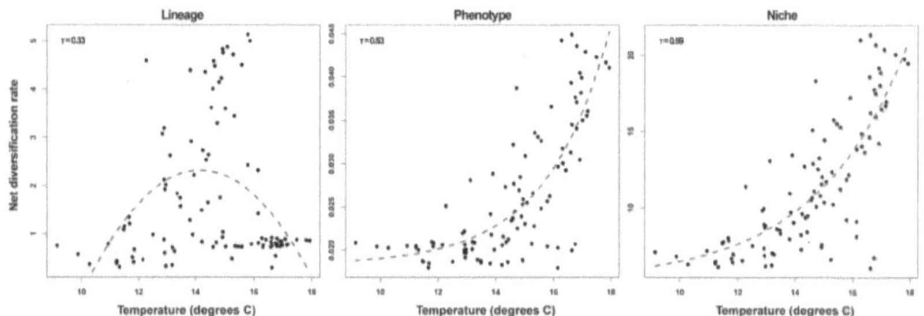

Global temperatures through time were inferred from climate model simulations of benthic delta-O-18 (de Boer *et al.*, 2014).

Figure 14. Best regressions of global temperature with diversification rates.

Chapter 4. Testing introgression hypotheses in *Penstemon* subgenus *Dasanthera*

Introduction

The evolutionary significance of introgression – the movement of genetic material between species – has been a contentious topic to evolutionary biologists for some time. Foundational work stemming from the modern synthesis popularized the concept of species as reproductively isolated populations (*e.g.,* Mayr, 1942; Stebbins, 1950). This focus on reproductive isolation as a primary mechanism of speciation led to a general tendency, at least from zoologists, to assume that introgression was an uncommon and unimportant driver of evolutionary change (Mayr, 1942; Barton, 2001; Coyne & Orr, 2004). Despite a plethora of evidence implicating a significant role for introgression in the generation of biological diversity in plants (Anderson, 1949; Stebbins, 1950; Anderson & Stebbins, 1954; Grant, 1981; Rieseberg & Wendel, 1993), the idea that this process could be both common and important in other taxonomic groups was largely dismissed until the advent of modern genomics (Mallett, 2007). Now, driven largely by the development of tools to detect and quantify genomic signatures of introgression, evidence of the historical and contemporary effects of introgression are apparent in the genomes of species across the tree of life, and introgression is increasingly recognized as common and taxonomically widespread (Mallett *et al.,* 2016; Taylor & Larson, 2019). This recognition has in turn invigorated the development of hypotheses regarding the genomic outcomes of introgression (*e.g.,* Schumer *et al.,* 2018; Hamlin *et al.,* 2020) and the role of introgression in generating biological diversity (Arnold & Kunte, 2017).

As a diversity-generating mechanism, introgression acts most profoundly, perhaps, through facilitating adaptive radiation: the proliferation of ecological roles and adaptations in different species within a lineage (Givnish, 1997). The main mechanistic link between the two is that hybridization between distantly related species can increase phenotypic and genetic variance (Seehausen, 2004; Givnish, 2010). This can result in, for example, the evolution of novel phenotypes through transgressive segregation, which can in turn stimulate the onset of adaptive radiation (Bell & Travis, 2005; Kagawa & Takimoto, 2018). And, while the importance of introgression in the context of adaptive radiation has, historically, been controversial (Seehausen, 2004), advances in our ability to identify the source, timing, and function of introgressed genetic material should continue to elucidate the link between the two (Marques *et al.,* 2019; Gillespie *et al.,* 2020). Indeed, evidence of introgression generating phenotypic and genetic diversity has been reported in several studies of adaptive radiations, across a diverse array of taxonomic groups, including *Heliconius* butterflies (Edelman *et al.,* 2019), Darwin's finches (Grant & Grant, 2019), cichlids (Svardal *et al.,* 2019), and wild tomatoes (Pease *et al.,* 2016). But perhaps the most prominent display of introgression as a diversifying mechanism is in the rapidly radiating genus *Penstemon.*

Penstemon Schmidel (Plantaginaceae), with nearly 300 described species, is the largest angiosperm genus endemic to North America (Wolfe *et al.,* 2006; Freeman, 2019). The exceptional array of floral diversity in *Penstemon* implicates a significant role of pollinator selective pressure (Pennell, 1935; Straw, 1956a, 1966), and, in tandem with lineage diversification patterns consistent with that of a rapid evolutionary radiation (Wolfe *et al.*, 2006; Wessinger *et al.*, 2016; Wolfe *et al.*, 2021), suggests that *Penstemon* represents an adaptive radiation. The occurrence of hybridization between species of *Penstemon* has been documented extensively, and this has fueled speculation about the relationship between introgression and speciation in the genus. For example, hybridization has been linked to the occurrence of whole-genome duplication in several *Penstemon* species (*e.g.*, Crosswhite & Kawano, 1965; Freeman, 1983; Broderick *et al.*, 2011), supporting early molecular studies which suggested polyploidization, and more generally, introgression, as potentially common mechanisms of speciation in the genus (Clausen *et al.*, 1940; Keck, 1932, 1945; Straw, 1955). Phylogenetic inquiries into *Penstemon* have also identified introgression as a key evolutionary process, because it likely explains, at least in part, the high degree of discordance observed between gene trees (Wolfe *et al.*, 2006, 2021; Wessinger *et al.*, 2016). However, the role of introgression as a diversifying mechanism in *Penstemon* has most often been discussed in the context of pollen-mediated gene flow, particularly between hymenopteran- and hummingbird-pollinated species. For example, Straw (1955, 1956a, 1956b) proposed two hypotheses of speciation derived from hybridization between hummingbird-pollinated and hymenopteran-pollinated species in cismontane southern California. Straw hypothesized that *P. spectabilis* (primarily hymenopteran) was formed through hybridization between *P. centranthifolius* (hummingbird) and *P. grinnellii* (hymenopteran), and that *P. clevelandii* (primarily hummingbird) was formed through subsequent backcrossing of *P. spectabilis* with *P. centranthifolius*. The mechanism

underlying these putative hybrid speciation events was that the stark differences in floral

morphologies between the hummingbird- and hymenopteran-pollinated served as a

reproductive isolating mechanism; Straw hypothesized that despite the apparent absence

of gametic barriers to interfertility, pollinator preferences made hybridization events

between these species rare, ultimately resulting in two diploid hybrid speciation events

(Straw, 1955, 1956a, 1956b).

While a hybrid origin was subsequently supported for only one of these proposed

speciation events (Wolfe *et al.*, 1998a, 1998b), Straw's hypotheses nevertheless reflect a

tendency of many *Penstemon* researchers to focus on the interplay between hybridization,

speciation, and floral evolution, especially as it pertains to adaptations for hummingbird

pollination. Other similar hypotheses involving hybridization between hymenopteran-

and hummingbird-pollinated species have been proposed (*e.g., P. xjonesii*) and

subsequently supported with molecular evidence (Crosswhite, 1967; Crump *et al.*, 2020).

Likewise, much research has been done on the genetic basis for transitions to

hummingbird pollination (*e.g.,* Wessinger & Rausher, 2013; Wessinger *et al.*, 2014) and

the differentiation of pollination traits in *Penstemon* (*e.g.,* Wilson & Jordan, 2009; Katzer

et al., 2019; Cardona *et al.*, 2020). From an evolutionary perspective, this predilection

with hummingbird pollination is warranted: transitions from insect pollination to

hummingbird pollination are estimated to have occurred 12-20 times in *Penstemon*

(Wilson *et al.*, 2007). Yet, such a heavy focus on hybridization as it relates to

hummingbird pollination may disproportionately emphasize the importance of this process in generating diversity; hummingbird-adapted species of *Penstemon* are relatively rare, and they diversify much more slowly than insect-pollinated species (Wessinger *et al.,* 2019). In this light, assessing the role of hybridization as a diversification mechanism in the absence of hummingbird pollination may help illuminate other processes that are integral to understanding speciation in *Penstemon*. Fortunately, an ideal group in which to address these questions can be found in *Penstemon* subgenus *Dasanthera*.

 Penstemon subgenus *Dasanthera* (Raf.) Pennell (hereafter, "*Dasanthera*") contains nine species distributed mainly in montane regions of the Pacific Northwest of North America. As currently circumscribed, *Dasanthera* comprises sixteen taxa: *P montanus* var. *montanus* and var. *idahoensis, P. ellipticus, P. lyallii, P. fruticosus* var. *fruticosus,* var. *scouleri,* and var. *serratus, P. davidsonii* var. *davidsonii,* var. *menziesii,* and var. *praeteritus, P. newberryi* var. *newberyii,* var. *berryi,* and var. *sonomensis, P. cardwellii, P. rupicola,* and *P. barrettiae.* The geographic distributions of *Dasanthera* species can be split broadly into two groups: those distributed primarily in the Cascades and Sierra Nevada Mountains (Figure 15), and those distributed primarily in the northern Rocky Mountains (Figure 16). Hybridization is common in *Dasanthera,* evidenced by the abundance of hybrid cultivars available (Vesall, 1990), the exceptional inter-species fertility observed in crossing experiments (Every, 1977), and the number of well-documented hybrid populations in nature (Clausen et al. 1940; Datwyler, 2001; Every, 1977). Frequent hybridization between species can provide a major obstacle for studies employing molecular sequence data, and this was likely a major barrier to early molecular phylogenetic studies in *Dasanthera* (Datwyler & Wolfe, 2004).

Phylogenetically, *Dasanthera* is sister to the rest of *Penstemon* (Wolfe et al. 2006), and

while recent work has elucidated the evolutionary relationships among *Dasanthera*

species in the Cascades and Sierra Nevada Mountains (Stone & Wolfe, 2020), several

taxonomic questions persist.

In general, floral morphology is relatively conserved across *Dasanthera* species

(Figure 17); members are characterized by densely pubescent anthers, strongly bilabiate

flowers typically between 30-40 mm in length, and a prominent keel which extends along

the entire adaxial surface of the corolla (Every, 1977; Datwyler, 2001; Freeman, 2019).

Two *Dasanthera* species (*P. newberryi* and *P. rupicola*) do exhibit some morphological

characteristics indicative of adaptation to hummingbird pollination (*i.e.,* pink corollas,

exserted stamens), and these species have indeed been the focus of studies attempting to

identify the sources of selection shaping patterns of introgression with *P. davidsonii*, a

non-hummingbird-adapted species (Clausen *et al.,* 1940; Every, 1977; Datwyler, 2001;

Kimball, 2008; Kimball *et al.,* 2008; Kimball & Campbell, 2009). However, the majority

of *Dasanthera* species, which have purple or blueish-purple flowers and anthers inserted

within the corolla, are likely visited most frequently by large hymenopterans (Every,

1977; Datwyler, 2001; Freeman, 2019). Many of these species hybridize when in

sympatry, potentially resulting in the formation of several taxa in the absence of

hummingbird pollination entirely. A. D. Every, in his 1977 book on the biosystematics of

Dasanthera, proposed that *P. fruticosus* var. *serratus* and *P. davidsonii* var. *praeteritus*

are the result of hybridization between *P. fruticosus* and *P. davidsonii*

(Every, 1977). Every provided support for these hypotheses in the late 1970's through the analysis of putative parent and hybrid chromatograph profiles (Every, 1977). Yet, despite remarkable advances in the molecular sequencing techniques and methodological approaches used to identify hybrid taxa since Every's work, these new tools have not yet been applied in a formal test of Every's hypotheses.

The goals of this study, therefore, are to employ next-generation sequence data to (1) test the hypothesis that hybridization between *P. davidsonii* and *P. fruticosus* resulted in the formation of varietal taxa for these species, and (2) resolve phylogenetic relationships in *Dasanthera*, including more taxa from the Northern Rocky Mountains than present in the most recent phylogeny of the subgenus (Stone & Wolfe, 2020). In achieving these goals, this study will resolve species' relationships in a group that has troubled *Penstemon* taxonomists for some time, and will contribute to our growing understanding of the role of introgression as a diversification mechanism in a group of plants which has undergone a recent adaptive radiation.

Materials and Methods

Data generation

We collected a total of 165 samples representing every species of *Dasanthera* except the rare and narrowly endemic *P. barrettiae*, which is found only along a narrow stretch of the Columbia River in Washington and Oregon. At the subspecific level, we were also unable to include samples of *P. montanus* var. *idahoensis*, which is endemic to central-west Idaho at high elevations (to 8000 feet), and *P. newberryi* var. *sonomensis*, which is a rare variety found only in the southernmost extent of the northern coast range in California. Our sampling represents 118 unique localities from across the geographic distributions of most species (Figures 16 and 17), and with intraspecific sampling focused most heavily on *P. davidsonii* and *P. fruticosus*, represented by 49 and 66 individuals, respectively (Table 18). Additionally, we included three species from subgenus *Penstemon* section *Caespitosi* (*P. abietinus*, *P. crandallii*, and *P. thompsoniae*) and one from subgenus *Penstemon* section *Cristati* (*P. acaulis*) for use as outgroup taxa.

DNA was extracted using a modified CTAB protocol (Wolfe, 2005) either from silica-dried leaf tissues (140 samples) or leaf tissues procured from specimens loaned from the University of Washington and Oregon State University herbaria (25 samples). We then generated high-throughput sequence data with a single-enzyme digest approach to Genotyping by Sequencing (GBS), as described in Elshire *et al.* (2011). GBS libraries were prepared with 100 nanograms of DNA per sample and via a modified version of the Elshire *et al.* (2011) protocol, described in Smith & Carstens (2020). GBS libraries were sequenced on an Illumina Hi-Seq 2500 using paired-end 150 bp sequencing at Novogene Corporation Inc. (Sacramento, CA). We assessed raw read quality with *fastQC* (Andrews, 2010); samples that generated substantially fewer reads (<250,000) were discarded from further analysis (and are not included in total counts of samples described above). We then processed forward reads in *ipyrad* (Eaton & Overcast, 2020). Raw reads were trimmed to 75 bp, and the maximum sampling depth per locus was limited to 100,000 reads. All other parameters for data processing were kept at default values, except that we altered the minimum number of samples per locus (parameter 21) and the samples included in the final assembly on an analysis-by-analysis basis (details in methods below). Summary statistics for each generated data set are listed in Table 19.

Phylogenetic inference

We inferred both a lineage tree and a species tree under a multispecies coalescent model with *SVDQuartets* (Chifman & Kubatko, 2014). For both analyses, the input data included all 165 *Dasanthera* samples, the four outgroup species, and required that loci were present in at least 25% of all included samples. Both analyses also evaluated all possible quartets and performed 100 bootstrap replicates. For the lineage tree, samples were not partitioned, such that each tip of the tree represents an individual sample. For the species tree, samples were partitioned by taxon, including at the subspecific level, resulting in 14 *Dasanthera* taxa, 3 taxa from Section *Caespitosi*, and 1 taxon from Section *Cristati*.

Hypothesis-tests of hybrid formation

We used the software *HyDe* (Blischak *et al.*, 2018) to test hypotheses of hybrid origins in varieties of *P. davidsonii* and *P. fruticosus*. For this analysis, we used a data set containing only *P. davidsonii*, *P. fruticosus*, and *P. abietinus*, and set the minimum number of samples per locus equal to 25% of the total number of included samples. To perform hypothesis tests, *HyDe* estimates the test statistic γ, which ranges from 0-1, and represents the proportion of parent ancestry in a taxon; for example, in a 50:50 hybrid, γ = 0.5. The test statistic γ, in conjunction with an assessment of statistical significance, enables *HyDe* to determine whether there is support for the hybrid formation of taxa, and to estimate the proportion of genetic contribution from each parent into the hybrid. We used *HyDe* to test whether there is molecular support for the putative hybrid origins of *P. davidsonii* var. *praeteritus* and *P. fruticosus* var. *serratus*, as outlined by Every (1977), but also to test support for the hybrid origins of *P. davidsonii* var. *menziesii* and *P. fruticosus* var. *scouleri*. In order to determine whether only some vs. all individuals display evidence of hybrid origins, we subsequently implemented individual *HyDe* analyses for each individual sample of these taxa.

Admixture network inference

We used the software *Treemix* (Pickrell & Pritchard, 2012) to infer an admixture network of the clade containing *P. fruticosus* and *P. davidsonii*, with *P. abietinus* as an outgroup. For this analysis, we used a data set which included the three aforementioned species, and required only four samples per locus in *ipyrad*, but restricted *Treemix* to infer admixture networks only from unlinked SNPs present in at least 25% of samples for each taxon. *Treemix* uses a maximum likelihood algorithm to estimate a population graph, including branch lengths (proportional to the amount of genetic drift between populations), and allowing for migration between populations. Model fit is assessed via the degree to which migration edges reduce the residual genetic covariance among populations. To infer the best model, we tested models containing 0-5 migration edges, performing five replicates each. For each replicate analysis, a random seed was generated to randomize the set of unlinked SNPs included in the analysis (in cases where multiple

SNPs occur in a single locus), but this seed was maintained across tests of different

migration edges to ensure model comparability (*i.e.,* the same random seed was used per

replicate across models implementing different numbers of migration edges). Model fit

was then compared across models of different migration edges and replicate data sets,

with the best model determined using the Akaike Information Criterion (AIC: Akaike,

1973).

Results

Phylogenetic inference

The species tree supports the monophyly of *Dasanthera* with a high degree of

support (Figure 18). Overall, species' relationships in the species tree largely reflect the

results observed in Stone & Wolfe (2020). However, the addition in this study of *P.

lyallii*, *P. ellipticus*, and several varieties of *P. davidsonii* and *P. fruticosus* add clarity to

the phylogenetic history of *Dasanthera* as a whole. Interestingly, *P. davidsonii* var.

praeteritus is nested within a clade containing *P. fruticosus* and its varieties, and this

relationship is supported strongly (100% of bootstrap replicates). Because *P. davidsonii*

and *P. fruticosus* are inferred as sister taxa here, and the close relationship between these

two species is evident, one might expect that if *P. davidsonii* var. *praeteritus* is indeed a

hybrid, its phylogenetic placement within *P. fruticosus* would not be wholly unexpected.

However, this placement of *P. davidsonii* var. *praeteritus* is also evident in the most

recent phylogenetic analysis of *Penstemon* (Wolfe *et al.,* 2021), suggesting this taxon

may be more accurately described as a variety of *P. fruticosus*. The phylogenetic placement of *P. montanus*, *P. lyallii*, and *P. ellipticus* here is also unclear; while the majority of bootstrap replicates suggest *P. montanus* is sister to the rest of the *Dasanthera* species (Figure 18), the remainder of bootstrap replicates suggest that either (1) *P. montanus* is sister to *P. lyallii* and *P. ellipticus* in a clade which is in turn sister to the rest of *Dasanthera*, or (2) *P. lyallii* and *P. ellipticus* are sister to all other *Dasanthera*, with *P. montanus* as the earliest diverging lineage therein.

Interspecific relationships in the lineage tree largely reflects those in the species tree, but additional insights are apparent with respect to intraspecific relationships. First, the lineage tree indicates that *P. cardwellii* is paraphyletic (Figure 19); this result was also observed in Stone & Wolfe (2020). Second, there is mixed evidence for the monophyly of species' varieties. While in most cases monphyly is supported (*i.e.*, *P. davidsonii* var. *davidsonii* and var. *praeteritus*, *P. fruticosus* var. *scouleri* and var. *serratus*, and both varieties of *P. newberryi*), in two others, it is not (*P. davidsoniii* var. *menziesii* and *P. fruticosus* var. *fruticosus*). This apparent monophyly of species' varieties likely reflects the strong influence of geography on population-genetic structure rather than any inherent mechanisms promoting reproductive isolation; most varieties have limited geographic distributions and do not display any reproductive barriers with conspecifics (Every, 1977).

Hypothesis-tests of hybrid formation

Of the four taxa tested for hybrid origins, two (*P. davidsonii* var. *menziesii* and *P. fruticosus* var. *serratus*) produced statistically significant results in the full *HyDe* analysis

(Table 20). Estimates of γ suggest that *P. davidsonii* var. *menziesii* is likely the product of hybrid speciation, as γ is very close to 0.5 (*p*-value $\ll$ 0.005). Conversely, *P. fruticosus* var. *serratus* is likely the result of hybridization followed by backcrossing with *P. fruticosus* var. *fruticosus*, as the estimate of γ for this taxon is not close to 0.5 (γ = ~0.25; *p*-value $\ll$ 0.005). Interestingly, *P. davidsonii* var. *praeteritus* shows no evidence of a hybrid origin (nonsensical γ estimate), and while *P. fruticosus* var. *scouleri* shows some evidence of introgression (γ = ~0.1), this result is not deemed statistically significant after a Bonferroni correction.

The individual *HyDe* tests identified at least one hybrid individual for each taxon (Table 20). However, the only taxon for which every individual was identified as a hybrid was *P. fruticosus* var. *serratus*, and the sampling for this taxon is limited (3 individuals). Within taxa, mean estimates of γ for the taxon as a whole tend to vary substantially (Mean γ population, Table 20), but when comparing only individuals identified as hybrids, the variation in estimates of γ is quite small (Mean γ hybrids, Table 20). This suggests that while admixture may not be uniform across all populations of a given taxon (except *P. fruticosusi* var. *serratus*), it is indeed uniform among the individuals identified as admixed.

Admixture network inference

The topology of the best *Treemix* graph is similar to the clade containing *P. fruticosus* and *P. davidsonii* in the species tree; a close relationship between *P. davidsonii* var. *davidsonii* and var. *menziesii* is apparent, and *P. davidsonii* var. *praeteritus* is nested within the *P. fruticosus* subclade, suggesting a close relationship with *P. fruticosus* var. *fruticosus* (Figure 20). This graph also includes one migration edge, which shows admixture from *P. davidsonii* into *P. fruticosus* var. *fruticosus* (migration weight w = 21%). However, using AIC, this model was only marginally better than the model with two migration edges, which added an admixture event from *P. abietinus* (outgroup) into *P. davidsonii* var. *davidsonii* (Table 21). There have been no documented cases of *in situ* gene flow between *P. davidsonii* and *P. abietinus*, and there is no overlap in their geographic distributions. These two species are in entirely different subgenera: *P. davidsonii* in subgenus *Dasanthera*, and *P. abietinus* in subgenus *Penstemon*. We know of no instances where hybridization between *Penstemon* species in different subgenera has been documented in nature. Furthermore, the migration weight for this second edge is very low (w = 1.5%). Given this, while we cannot rule out the possibility of low amounts of admixture between *P. abietinus* and *P. davidsonii*, we find this result unlikely, and suspect it may reflect some historical process affecting the segregation of ancestral alleles during speciation (*e.g.,* incomplete lineage sorting).

Discussion

On the hybrid origins of P. fruticosus and P. davidsonii varieties

Although the first to hypothesize hybrid origins of *P. davidsonii* var. *praeteritus* and *P. fruticosus* var. *serratus* was Every (1977), morphological similarities between *P. davidsonii* var. *praeteritus* and *P. fruticosus* were noted when the taxon was first described (Cronquist, 1964). Both hypothesized hybrid taxa are morphologically intermediate to their hypothesized parent taxa in characters such as the ratio of leaf length to leaf width (smaller in *P. davidsonii*), the size of the inflorescence (fewer flower-bearing nodes in *P. davidsonii*), the type of leaf margin (serrate or dentate in *P. fruticosus*, entire in *P. davidsonii*), and the length of the corolla (shorter in *P. davidsonii*) (Cronquist, 1964; Every, 1977; Freeman, 2019). Every (1977) suggested that glacial dynamics during the Pleistocene could have served as a mechanism to promote distributional overlap and hybridization between parent species in eastern Oregon, where both putative hybrid taxa are found (but only one putative parent: *P. fruticosus*). Every (1977) hypothesized that the cooler climatic conditions of the Pleistocene would have provided suitable habitat for *P. fruticosus* and *P. davidsonii* at lower elevations in the Columbia Basin and Intermountain Region, where they would be in close contact and could hybridize. When the glaciers began to retreat, these areas would once again become inhospitable, leaving only 'islands' of suitable habitat in relatively isolated outposts at higher elevations. Populations that persisted in these areas would likely have both *P. fruticosus* and *P. davidsonii* ancestry. If supported, Every's hypotheses would explain the disjunct distribution of *P. davidsonii* var. *praeteritus*, which is found only in the Steens Mountains in southeastern Oregon and the Black Rock Range in northwestern Nevada, far from the Cascades and Sierra Nevada ranges where *P. davidsonii* is typically found. Likewise, it would explain the endemism of *P. fruticosus* var. *serratus*, found only in the Blue and Wallowa and Seven Devils Mountains of southeastern Washington, northeastern Oregon, and west-central Idaho.

We found mixed support for the hypothesized hybrid origins of *P. fruticosus* var. *serratus* and *P. davidsonii* var. *praeteritus* outlined by Every (1977). *HyDe* detected evidence of introgression in *P. fruticosus* var. *serratus*, but failed to detect any signal of hybridization in *P. davidsonii* var. *praeteritus* (Table 20). Despite the lack of support for a hybrid origin of *P. davidsoniii* var. *praeteritus*, evidence of introgression in *P. fruticosus* var. *serratus* lends credence to Every's emphasis on the role of glaciation in altering species' distributions in *Dasanthera*. Likewise, species distribution models of *P. davidsonii* and *P. fruticosus* during the Last Glacial Maximum suggest that there would indeed have been ample suitable habitat for both of these species in the Columbia Basin and Intermountain Region (Stone & Wolfe, 2020; Figure 8). Following glacial retreat, populations in these regions would have been separated from the rest of the contiguous distribution of their parent species. We agree that this scenario is the most likely one to explain the introgression observed in *P. fruticosus* var. *serratus*, and we find no other reasonable explanation for the disjunct distribution of *P. davidsonii* var. *praeteritus*. It is somewhat curious, then, that phylogenetic analyses place this taxon within the *P. fruticosus* clade (Figure 18; Figure 19), and that our hybridization tests and admixture analyses suggest no signal of hybridization in this taxon (Table 20; Figure 20). We doubt that this result is due to the failure of *HyDe* to detect admixture which is truly present, as *HyDe* retains high statistical power to detect hybridization (without having an overly high false discovery rate) and gives robust estimates of γ even when its assumptions about hybridization are violated (Kong & Kubatko, 2021). Furthermore, the identification of introgressed genetic material in *P. fruticosus* var. *serratus* and *P. davidsoniii* var.

104

menziesii confirms a history of hybridization between these species, a phenomenon that has been observed in natural populations and reported in previous molecular studies (Every, 1977; Datwyler, 2001; Stone & Wolfe, 2020).

Instead, we interpret these results as evidence that *P. davidsonii* var. *praeteritus* would be more accurately described as a variety of *P. fruticosus*. In this case, the morphological affinity of this taxon to *P. davidsonii* var. *davidsonii* may be the result of phenotypic plasticity or adaptation to similar environmental conditions. *P. davidsonii* var. *praeteritus* is found mainly on rock outcrops and in crevices in the subalpine- to alpine zone, which begins at roughly 2500 meters on Steens Mountain (Mansfield, 1996), while *P. davidsonii* var. *davidsoniii* can occur as high as 3700 meters in the Cascades Mountains (Freeman, 2019). The evidence provided by Every to support the hybrid origin of this taxon relied on the intermediacy of morphological characters with *P. davidsonii* var. *davidsonii* and *P. fruticosus* var. *fruticosus* and the analysis of substances assumed to be flavonoid compounds via paper chromatography profiles (Every, 1977). We argue that the results generated from the analysis of sequence data in this study present a more compelling case, such that we reject the hypothesis of a hybrid origin for *P. davidsonii* var. *praeteritus* and suggest that this taxon would be more appropriately placed under *P. fruticosus*.

Phylogenetic relationships in Dasanthera

We inferred a species tree for *Dasanthera* which includes thirteen of the sixteen total taxa in the subgenus and has higher resolution than in any previous study. Early attempts to estimate relationships between *Dasanthera* species suffered from gene tree

incongruence and limited interspecific variation in gene sequences (Datwyler & Wolfe,

2004; Wolfe *et al.*, 2006). While a more recent phylogeny of *Dasanthera* did not suffer

from these issues, the goals of that study were primarily to estimate relationships between

species distributed in the Cascades and Sierra Nevada Mountains; as a result, several taxa

in the subgenus, including *P. davidsonii* var. *praeteritus*, *P. fruticosus* var. *serratus* and

var. *scouleri*, *P. lyallii*, and *P. ellipticus*, were not included (Stone & Wolfe, 2020).

The results presented here elucidate most of the relationships between *Dasanthera*

taxa with strong support (Figure 18; Figure 19). Our species tree shows a strong

geographic component to species' relationships, with a clear distinction between species

distributed primarily in the Cascades and Sierra Nevada Mountains vs. those distributed

primarily in the northern Rocky Mountains (Figure 18). The exception to this distinction

is *P. fruticosus*, which, perhaps unsurprisingly, shows a close relationship with *P.

davidsonii*. However, there is still uncertainty in the relative phylogenetic placement of *P.

ellipticus*, *P. lyallii*, and *P. montanus*. Bootstrap support for this branch in the species tree

is weak (61%), and the two alternative supported topologies, which received roughly

equal bootstrap support, suggested either that these three species form a clade, or that the

placement of *P. montanus* as sister to the rest of the subgenus should be swapped with the

clade containing *P. ellipticus* and *P. lyallii*. Hybridization is known to occur between *P.

ellipticus* and *P. lyallii*, and although these species (and other *Dasanthera* taxa) are likely

interfertile with *P. montanus*, hybridization with *P. montanus* has never been documented

(Datwyler, 2001). This includes an apparent lack of introgression with *P. fruticosus*,

despite extensive overlap in the geographic overlap of these two species (Figure 16).
Therefore, while it is possible that the uncertainty regarding the phylogenetic placement
of *P. montanus*, *P. ellipticus*, and *P. lyallii* is because the lone population of *P. montanus*
included has experienced introgression from another species, this seems unlikely. This
leaves introgression from *P. fruticosus* into *P. lyallii* and/or *P. ellipticus* as the most
likely source of uncertainty in this branch of the species tree. Future studies examining
introgression patterns in the northern Rocky Mountain taxa and denser sampling of *P.
montanus* are needed to clarify this uncertainty.

Introgression as a diversification mechanism in Penstemon

Our results provide evidence that introgression can serve as a mechanism for
diversification in *Penstemon*, even in the absence of hummingbird-mediated gene flow.
This evolutionary role of introgression despite the relatively conserved floral morphology
of *Dasanthera* species is in direct contrast with a historical emphasis in *Penstemon*
research on the role of floral evolution in generation species diversity (*e.g.,* Pennell,
1935; Straw, 1966). This concept is not necessarily novel in *Penstemon*. Hybridization
has undoubtedly been critical to the diversification of Section *Proceri*, whose species
exhibit little variation in floral morphology (Keck, 1932; Blischak *et al.,* 2020). However,
much of the diversification in Section *Proceri* has been associated with the
polyploidization of taxa. While this certainly distinguishes evolution in Section *Proceri*
from speciation via hummingbird-mediated gene flow, it also differs from what has
occurred in *Dasanthera*; there are no known polyploid *Dasanthera* species (Clausen *et*

al., 1940; Broderick *et al.,* 2011; Freeman, 2019). This distinction brings into question the means by which introgression could be functioning as a diversification mechanism in *Dasanthera.*

On one hand, the process of adaptive introgression, whereby introgressed alleles are favored by natural selection (Gonzalez *et al.,* 2018), should favor the movement of specific alleles across species boundaries. Adaptive introgression has been inferred to be a diversification mechanism across a broad range of taxa (*e.g.,* Pease *et al.,* 2016; Edelman *et al.,* 2019; Schumer et al., 2018). Conversely, introgression can be opposed by selection, including through the accumulation of genetic incompatibilities between the genomes of parent species (Orr, 1995), or by extrinsic forms of selection. Between these two extremes, the degree to which alleles are exchanged between species depends on several other factors, including their genomic location and their proximity to loci under selection. For example, if introgressed loci are underrepresented in gene-rich regions of the genome, or in low recombination regions of the genome, this may suggest that alleles linked to incompatibility loci have been selectively purged from hybrid populations (Edelman *et al.,* 2019). Assessing how introgression shapes diversification dynamics in *Dasanthera* would thus depend on (1) the identification of major sources of selection related to the occurrence of hybridization when species' distributions overlap, and (2) understanding the functional importance of genomic regions responsible for phenotypic variation and fitness (Suarez-Gonzalez et al., 2018; Taylor & Larson, 2019). Aside from pollinator differences, studies on hybrid zones between *P. davidsonii* and *P. newberryi* have implicated elevation partitioning as a source of selection shaping introgression patterns (Kimball, 2008; Kimball et al., 2008). An elevational gradient in water availability appears to explain high hybrid fitness at intermediate elevations, but reduced fitness at more extreme elevations (Kimball & Campbell, 2009). Similarly, elevation partitioning also appears to function as a selective isolating mechanism in hybrid zones between *P. davidsoniii* and *P. rupicola* (Every, 1977; Datwyler, 2001). Future studies pairing *in situ* genomic variation with direct experimental analysis of traits related to fitness along elevational and water-availability gradients would therefore contribute greatly to our understanding of the relationship between introgression dynamics and diversification in both *Dasanthera* and *Penstemon*.

Tables and Figures

Table 18. Collection information for specimens included in Chapter 4.

Taxon	Latitude	Longitude	Population ID	# of Individuals
P. cardwellii	44.274483	-123.579297	cardwellii_BWS_91	1
P. cardwellii	45.210899	-123.758863	cardwellii_BWS_92	1
P. cardwellii	44.808767	-122.110374	cardwellii_BWS_108-5-10	1
P. cardwellii	44.401918	-121.860045	cardwellii_BWS_109	1
P. cardwellii	44.024151	-121.804755	cardwellii_BWS_110	1
P. cardwellii	46.313721	-122.036324	cardwellii_BWS_28	1
P. cardwellii	42.626077	-123.835088	cardwellii_BWS_90	1
P. cardwellii	45.417722	-121.825086	cardwellii_BWS_96	1
P. cardwellii	46.29557	-122.352124	cardwellii_BWS_97	1
P. dav. var. dav.	37.188755	-118.591939	dav. Wilson 3554	1
P. dav. var. dav.	44.1875	-121.88055	dav. SLD 39	1
P. dav. var. dav.	42.944241	-122.176547	dav._BWS_115	2
P. dav. var. dav.	42.895602	-122.092252	dav._BWS_116	2
P. dav. var. dav.	43.684535	-121.265441	dav._BWS_117	2
P. dav. var. dav.	44.266403	-121.787933	dav._BWS_23	3
P. dav. var. dav.	44.021183	-121.80011	dav._BWS_25	2
P. dav. var. dav.	44.023102	-121.775986	dav._BWS_27	3
P. dav. var. dav.	42.08102	-122.718583	dav._BWS_87	2
P. dav. var. dav.	45.394993	-121.658014	dav._BWS_95	2
P. dav. var. dav.	44.0917	-121.6467	dav._OSTL_1	1
P. dav. var. dav.	44.4861	-121.8386	dav._OSTL_2	1
P. dav. var. dav.	42.555233	-120.791667	dav._UWTL_11	1
P. dav. var. dav.	46.9125	-121.801667	dav._UWTL_13	1
P. dav. var. dav.	44.0052	-121.6424	dav._UWTL_3	1
P. dav. var. dav.	43.68781	-121.25558	dav._UWTL_4	1
P. dav. var. menz.	48.716888	-121.078214	menz._BWS_103	2
P. dav. var. menz.	48.706497	-121.13966	menz._BWS_104	1
P. dav. var. menz.	48.832448	-121.650946	menz._OSTL_4	1

Table 18

Table 18 continued

P. dav. var. menz.	48.734817	-120.675033	menz._UWTL_1	1
P. dav. var. menz.	47.50596	-123.2394	menz._UWTL_2	1
P. dav. var. menz.	47.391242	-121.62573	menz._UWTL_22	1
P. dav. var. menz.	48.52361	-120.81444	menz._UWTL_5	1
P. dav. var. menz.	48.9363	-120.20861	menz._UWTL_7	1
P. dav. var. menz.	48.902517	-121.466933	menz._UWTL_8	1
P. dav. var. praet.	-	-	praet._2_0471	1
P. dav. var. praet.	-	-	praet._96_0049	1
P. dav. var. praet.	-	-	praet._96_0105	1
P. dav. var. praet.	42.73539	-118.633691	praet._BWS_118	2
P. dav. var. praet.	42.720118	-118.626557	praet._BWS_119	2
P. dav. var. praet.	42.708274	-118.572401	praet._BWS_120	2
P. dav. var. praet.	42.667229	-118.582511	praet._BWS_121	2
P. dav. var. praet.	42.640212	-118.58301	praet._BWS_122	2
P. ellipticus	47.570698	-115.737701	ellipticus_BWS_41	1
P. ellipticus	48.374973	-116.158867	ellipticus_BWS_46	1
P. ellipticus	48.810383	-116.58847	ellipticus_BWS_48	1
P. ellipticus	48.798992	-116.609612	ellipticus_BWS_49	1
P. frut. var. frut.	43.040712	-116.691558	frut._Pop2	6
P. frut. var. frut.	43.295783	-115.328516	frut._Pop3	8
P. frut. var. frut.	43.357149	-115.447066	frut._Pop4	3
P. frut. var. frut.	-	-	frut._96_0023	1
P. frut. var. frut.	-	-	frut._96_0149	1
P. frut. var. frut.	47.411009	-121.092183	frut._BWS_102	2
P. frut. var. frut.	46.768161	-121.67451	frut._BWS_106-1-5	2
P. frut. var. frut.	46.906403	-121.578231	frut._BWS_107	2
P. frut. var. frut.	44.985237	-116.072357	frut._BWS_54	1
P. frut. var. frut.	45.229385	-115.794212	frut._BWS_55	2
P. frut. var. frut.	45.090702	-116.044273	frut._BWS_56	2
P. frut. var. frut.	43.753498	-114.28598	frut._BWS_58	2
P. frut. var. frut.	43.791691	-114.574608	frut._BWS_59	2
P. frut. var. frut.	44.383102	-114.724228	frut._BWS_60	2
P. frut. var. frut.	44.511154	-114.374054	frut._BWS_62	2
P. frut. var. frut.	44.46064	-113.296562	frut._BWS_63	2
P. frut. var. frut.	45.438129	-113.120415	frut._BWS_69	2

Table 18

Table 18 continued

P. frut. var. frut.	43.811768	-114.259209	frut._BWS_7	3
P. frut. var. frut.	45.42266	-121.614503	frut._BWS_93	1
P. frut. var. frut.	45.2569	-117.0875	frut._OSTL_11	1
P. frut. var. frut.	44.97	-118.2378	frut._OSTL_12	1
P. frut. var. frut.	45.3859	-117.4238	frut._OSTL_13	1
P. frut. var. frut.	44.284	-118.6798	frut._OSTL_14	1
P. frut. var. frut.	44.509	-120.6309	frut._OSTL_15	1
P. frut. var. frut.	44.236628	-115.212413	frut._OSTL_16	1
P. frut. var. frut.	45.312	-117.3064	frut._OSTL_9	1
P. frut. var. scoul.	48.587567	-120.42133	scoul._OSTL_17	1
P. frut. var. scoul.	48.594404	-118.655105	scoul._OSTL_19	1
P. frut. var. scoul.	48.828859	-118.831378	scoul._OSTL_20	1
P. frut. var. scoul.	48.731733	-118.513057	scoul._OSTL_21	1
P. frut. var. scoul.	48.612418	-118.481135	scoul._OSTL_22	1
P. frut. var. scoul.	48.151115	-116.860602	scoul._Pop7	2
P. frut. var. scoul.	48.2721	-117.6589	scoul._Popnew1	1
P. frut. var. scoul.	48.422159	-117.672191	scoul._Popnew2	2
P. frut. var. serr.	-	-	serr._2_047	1
P. frut. var. serr.	46.106752	-117.836277	serr._Pop8	2
P. mont. var. mont.	44.511154	-114.374046	mont._BWS_61	2
P. lyallii	47.50235	-115.881226	lyallii_BWS_33	1
P. lyallii	47.632248	-116.307442	lyallii_BWS_37	1
P. lyallii	47.698517	-116.376099	lyallii_BWS_38	1
P. lyallii	47.656219	-115.965439	lyallii_BWS_39	1
P. lyallii	47.74873	-115.972748	lyallii_BWS_42	1
P. lyallii	48.344994	-116.231789	lyallii_BWS_43	1
P. lyallii	48.371445	-116.156052	lyallii_BWS_44	1
P. lyallii	48.336964	-116.172638	lyallii_BWS_45	1
P. lyallii	47.479794	-116.060509	lyallii_BWS_50	1
P. lyallii	47.52282	-115.697975	lyallii_BWS_51	1
P. lyallii	47.522911	-115.697899	lyallii_BWS_52	1
P. lyallii	46.650951	-112.087273	lyallii_BWS_64	1
P. lyallii	47.411816	-112.714424	lyallii_BWS_66	1
P. newb. var. newb.	37.181157	-118.559254	newb._BWS_70	1

Table 18

Table 18 continued

P. newb. var. newb.	37.587992	-118.982346	newb._BWS_71	1
P. newb. var. newb.	38.021527	-119.26708	newb._BWS_72	1
P. newb. var. newb.	38.77672	-119.888744	newb._BWS_73	1
P. newb. var. newb.	38.615967	-119.915853	newb._BWS_74	1
P. newb. var. newb.	38.661708	-120.130671	newb._BWS_75	1
P. newb. var. newb.	38.97312	-120.094277	newb._BWS_76	1
P. newb. var. newb.	39.947764	-121.140946	newb._BWS_79	1
P. newb. var. newb.	39.880031	-121.161161	newb._BWS_80	1
P. newb. var. newb.	41.231111	-122.382305	newb._BWS_85	1
P. newb. var. berryi	40.658657	-123.219845	berryi_BWS_81	1
P. newb. var. berryi	40.585266	-123.542329	berryi_BWS_83	1
P. newb. var. berryi	41.389587	-122.993862	berryi_BWS_84	1
P. rupicola	47.411009	-121.092183	rupicola_BWS_101	1
P. rupicola	47.333277	-121.384662	rupicola_BWS_99	1
P. rupicola	47.268312	-121.366167	rupicola_BWS_100	1
P. rupicola	46.778033	-121.760309	rupicola_BWS_105	1
P. rupicola	44.024151	-121.804755	rupicola_BWS_111	1
P. rupicola	43.09002	-122.249755	rupicola_BWS_113	1
P. rupicola	41.223113	-122.37963	rupicola_BWS_86	1
P. rupicola	42.236884	-123.797584	rupicola_BWS_89	1
P. rupicola	47.436029	-121.778519	rupicola_BWS_98	1

Table 19. GBS data set statistics for each analysis.

Post-processing GBS data generation statistics for each analysis conducted in this study. LPS = average loci per sample; SNPs = total single nucleotide polymorphisms; H_e = average heterozygosity estimates.

Analysis	Total samples	Total loci	LPS	SNPs	% missing SNPs	H_e
Phylogenetic inference	169	2576	1565 ± 296	21806	36.90%	0.0123 ± 0.0071
HyDe	116	3094	1959 ± 342	20721	35.20%	0.0102 ± 0.0075
Treemix	116	69998	5883 ± 5374	303730	89.70%	0.0103 ± 0.0076

Table 20. Summary of *HyDe* tests: *P. davidsonii* var. *davidsonii* x *P. fruticosus* var. *fruticosus*.

All tests were performed with *P. davidsonii* var. *davidsonii* as parent 1, and P. *fruticosus* var. *fruticosus* as parent 2. Significant *p*-values are signified with an asterisk. The column "# of hybrid individuals" indicates the number of individuals from a population identified as hybrids for each test.

Putative Hybrid	Hypothesis tests			Individual tests		
	Z-score	*p-value*	γ	# of hybrid individuals	Mean γ hybrids	Mean γ population
P. davidsonii var. *praeteritus*	-9999	1	-0.083	3/13 (23.1%)	0.25 ± 0.02	-0.82 ± 4.85
P. davidsonii var. *menziesii*	5.96	*1.38 e-09* *	0.526	6/10 (60%)	0.44 ± 0.11	0.35 ± 0.18
P. fruticosus var. *serratus*	11.7	*0* *	0.249	3/3 (100%)	0.24 ± 0.06	0.24 ± 0.06
P. fruticosus var. *scouleri*	1.89	0.0295	0.102	4/10 (40%)	0.27 ± 0.03	0.18 ± 0.91

Table 21. *Treemix* results.

Five replicate data sets of randomly selected unlinked SNPs were used to test models
with different numbers of migration edges; each replicate data set was the same across
migration edges. Log-likelihood and ΔAIC values for each number of migration edges
are for the best model across replicate data sets.

Migration edges	logLik	ΔAIC
0	154.716	2.04
1	156.738	0.00
2	157.523	0.43
3	157.798	1.88
4	157.835	3.81
5	157.909	5.66

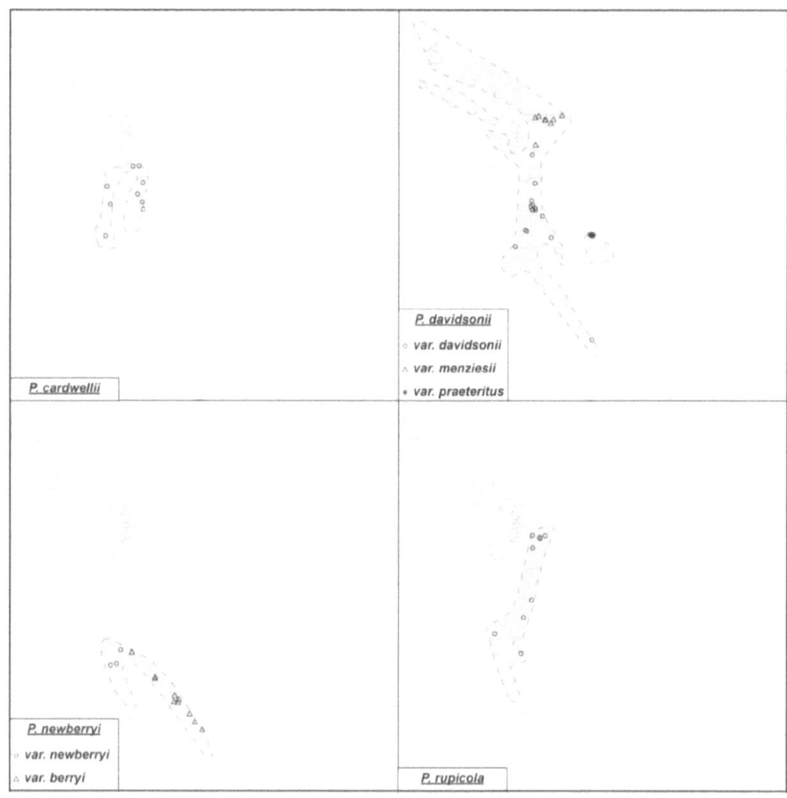

Geographic distributions and samples included in this study for species distributed primarily in the Cascades and Sierra Nevada Mountains (*P. cardwellii*, *P. davidsonii*, *P. newberryi*, and *P. rupicola*). Species' distributions are outlined in grey. Icons represent the sampling localities of individuals included in this study.

Figure 15. Range maps for Cascades and Sierra Nevada species.

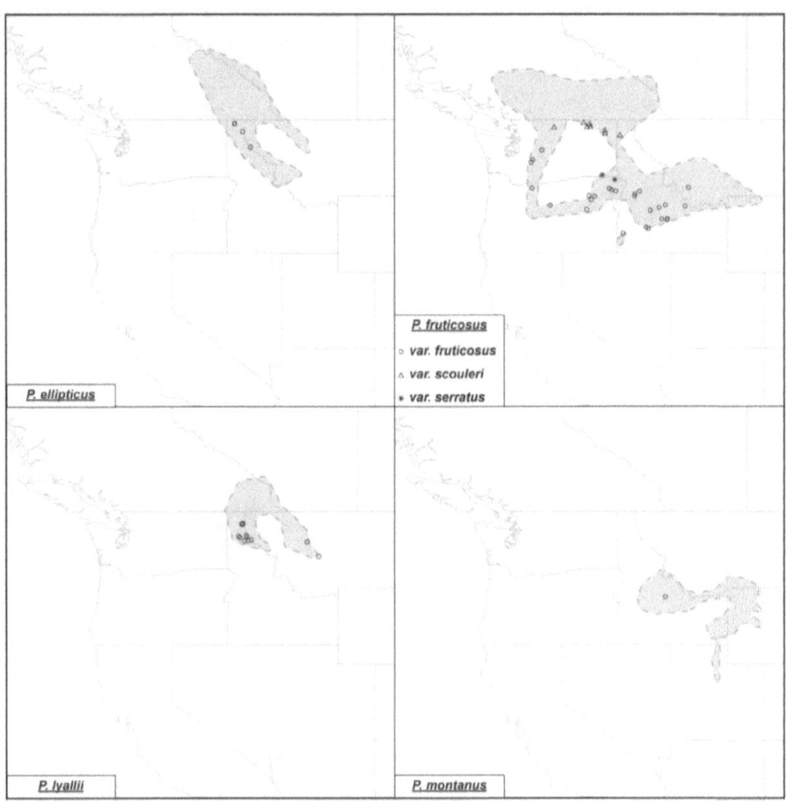

Geographic distributions and samples included in this study for species distributed primarily in the northern Rocky Mountains (*P. ellipticus*, *P. fruticosus*, *P. lyallii*, and *P. montanus*). Species' distributions are outlined in grey. Icons represent the sampling localities of individuals included in this study.

Figure 16. Range maps for northern Rocky Mountains species.

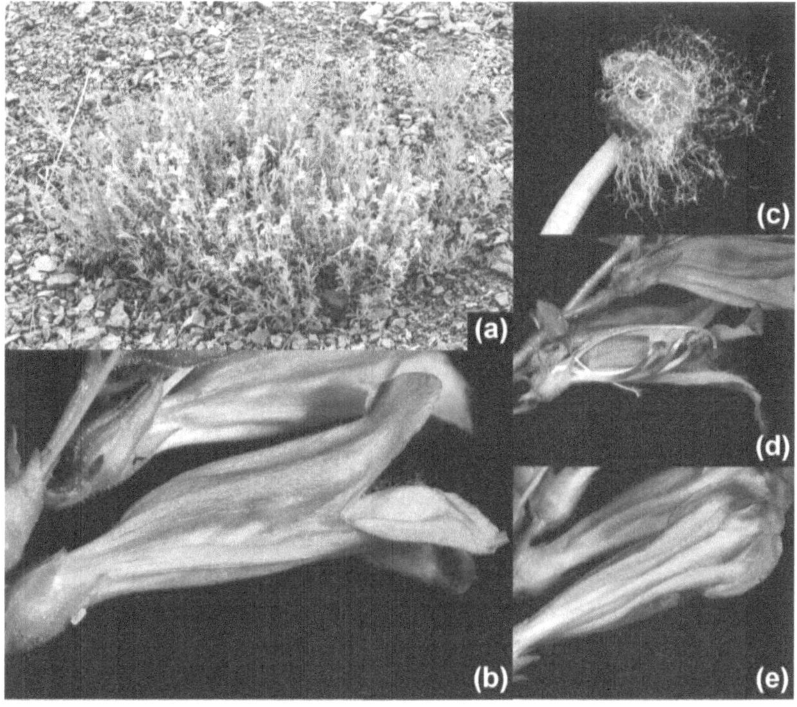

(a) Typical erect subshrub habit. (b) Profile of the corolla, with prominent ridge along the adaxial surface typical of the subgenus. (c) Anther covered with dense lanate hairs typical of the subgenus. (d) Dissected profile of the corolla exposing the staminode, stamens, style, and stigma. (e) Abaxial surface of the corolla showcasing prominent ridges. Photographs provided courtesy of Dr. Andi Wolfe.

Figure 17. Photographs of *Penstemon fruticosus*.

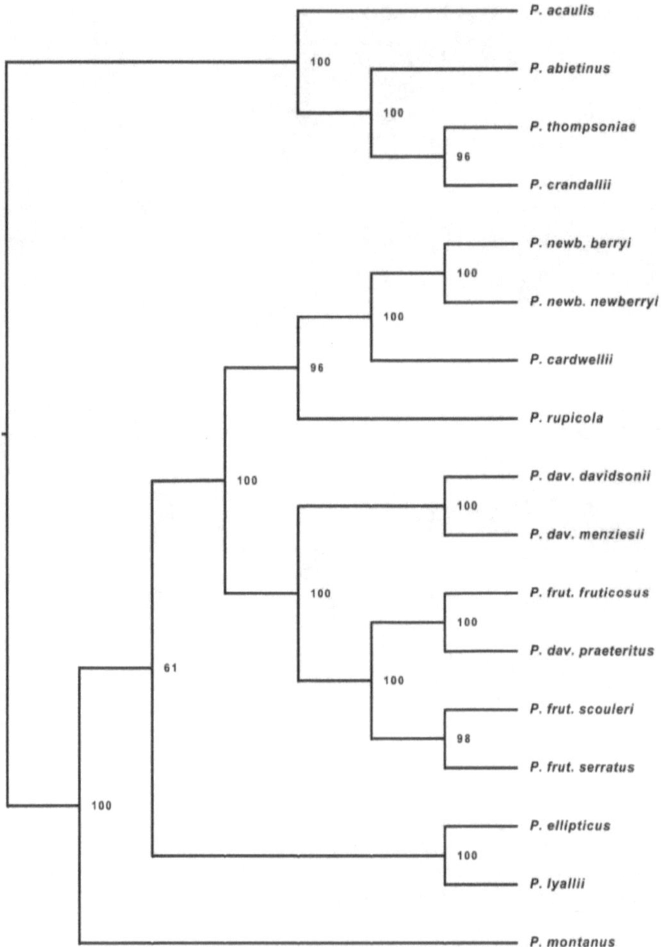

Values at the nodes indicate bootstrap support (100 bootstrap replicates).

Figure 18. Species tree constructed in *SVDQuartets*.

120

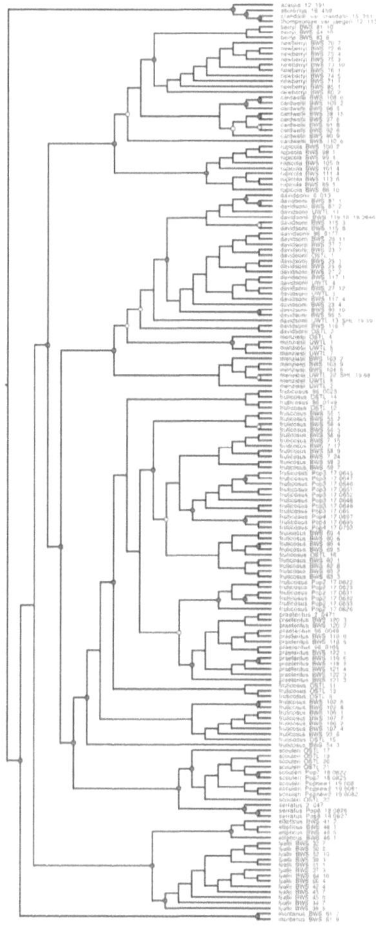

Icons at the nodes indicate the degree of bootstrap support; grey circle = 90-100%; white circle = 75-89%; no circle = < 75%.

Figure 19. Lineage tree constructed in *SVDQuartets*.

121

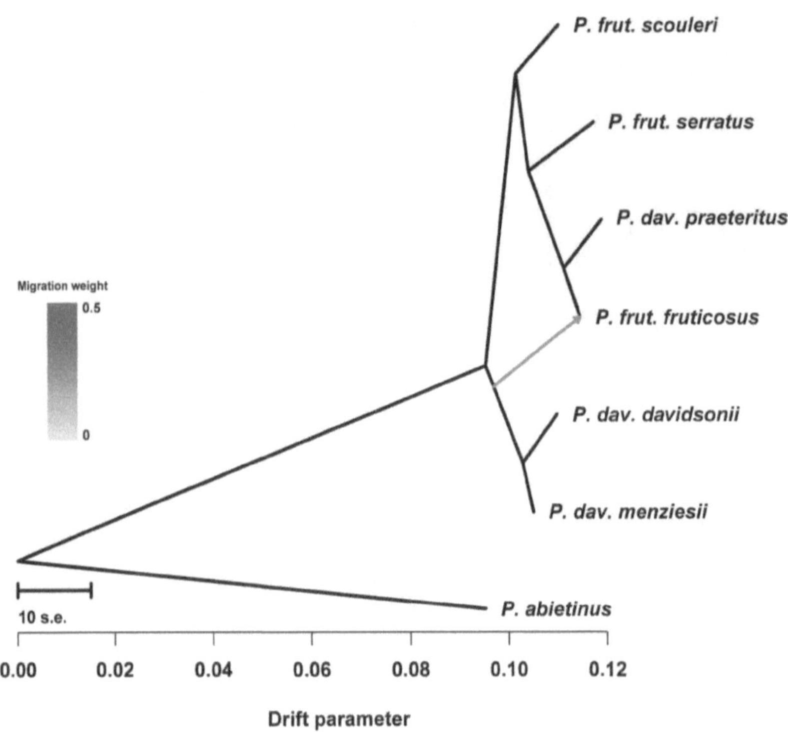

The scale bar shows ten times the average standard error of entries in the sample covariance matrix. The single estimated migration event is represented by an arrow, and is colored according to its migration weight.

Figure 20. Best population admixture graph inferred with *Treemix*.

122

Chapter 5. Conclusion

Overview

The aims of my book were to understand diversification in the genus *Penstemon*

at three taxonomic scales: across the entire genus, across species in *Penstemon* subgenus

Dasanthera, and within *Dasanthera* species. Across the genus, I found evidence of an

early burst of lineage diversification rates coincident with heightened glacial activity in

during the Pleistocene (Chapter 3). This burst was asynchronous with increases in

diversification rates for phenotypic evolution, contributing to a growing body of research

suggesting that asynchronicity in diversification rate shifts may be common, and

suggesting a primarily niche-neutral mode of diversification in the early stages of

Penstemon evolution. Across *Dasanthera* species, I continued to identify a critical role of

glaciation in patterns of diversification. Using species distribution modeling approaches, I

found that, contrary to the classical paradigm of temperate species' responses to

glaciation, *Dasanthera* species likely experienced increases in available suitable habitat

during the Last Glacial Maximum (Chapter 2). These distributional changes, in turn,

likely shaped introgression dynamics for species currently distributed in the Cascades and

Sierra Nevada Mountains, indicated by evidence linking the bulk of introgression events

in these taxa back to the Klamath Mountains. Within *Dasanthera* species, I identified

primarily north-to-south patterns of genetic

123

differentiation consistent with expectations for phylogeographic studies of taxa

distributed in the Pacific Northwest of North America (Chapter 2). In addition, I found

mixed support for classical hypotheses of hybrid taxon formation; in doing so, I

generated a strongly supported phylogeny of *Dasanthera*, clarifying relationships

between most of the taxa in the subgenus, and provided support for an updated taxonomic

circumscription of *P. davidsonii* and *P. fruticosus* (Chapter 4).

Future Directions

Taxonomic issues in Penstemon subgenus Dasanthera

While the work I have presented in this book has achieved most of my main aims,

many questions persist. Regarding the taxonomy of *Dasanthera*, species limits will likely

need to be revisited for *P. cardwellii*. This species, as currently circumscribed, was

shown to be paraphyletic (Chapter 2). This is most likely related to the apparent

propensity of this species to hybridize with its congeners, given the pervasive

introgression documented. Specifying the timing of this introgression, and, if possible,

identifying the ecological and geographical factors promoting it, would contribute greatly

to achieving a delimitation of this species which is consistent with phylogeny. To that

end, an implementation of phylogenetic methodologies which incorporate hybridization

under the multispecies coalescent would be a useful contribution. Continuing with

persistent taxonomic issues, I was unable to convincingly clarify relationships between

Dasanthera species at the base of the clade. While it appears clear that *P. lyallii* and *P.

ellipticus* are sister species, the relative placement of

these taxa in the *Dasanthera* phylogeny, along with *P. montanus*, is still uncertain. This is perhaps caused by introgression into *P. lyallii* and *P. ellipticus* from *P. fruticosus*, but support for this hypothesis is currently minimal. Future phylogenetic studies which focus on understanding the reason for this uncertainty are warranted. Finally, a well-supported phylogeny of *Dasanthera* which includes all described taxa has yet to be achieved. In my final estimate of the phylogeny, I was unable to include three taxa: *P. montanus* var. *idahoensis*, *P. barrettiae*, and *P. newberryi* var. *sonomensis*. Incorporating these taxa into future phylogenetic studies is undoubtedly necessary for a comprehensive interpretation of species limits in *Dasanthera*.

Processes governing diversification

Understanding the myriad processes governing the generation of biological diversity is of great interest to many biologists. In this sense, perhaps the most fruitful avenues for future research in *Penstemon* are those that more thoroughly examine processes affecting speciation, adaptation to different ecological niches, and the evolution of novel phenotypes. In particular, future studies should (1) employ comparative methodological approaches to examine the diversification of floral structure as it relates to likely pollinators, (2) examine the role of hybridization and introgression in both the generation of new species and the maintenance of existing species boundaries, (3) consider the geography of speciation, with special attention to range asymmetries and the overlap in ecological niche space, and (4) identify the genomic bases for variation in traits linked to adaptation and speciation. Futures studies in these four areas would address gaps in our understanding of how diversification has progressed in *Penstemon*, and are likely to meaningfully contribute to our growing understanding of this wonderful genus.

www.ingramcontent.com/pod-product-compliance
Lightning Source LLC
LaVergne TN
LVHW040320200726
843493LV00015B/1822